LIGHTING EFFICIENCY APPLICATIONS

LIGHTING

EFFICIENCY APPLICATIONS

Second Edition

Albert Thumann, P.E., CEM

Published by
THE FAIRMONT PRESS, INC.
700 Indian Trail
Lilburn, GA 30247

Library of Congress Cataloging-in-Publication Data

Lighting efficiency applications / [compiled by] Albert Thumann. --
2nd ed.
    p. cm.
    Includes bibliographical references and index.
    ISBN 0-88173-137-4
    1. Electric lighting. 2. Electric lighting--Energy conservation.
I. Thumann, Albert.
TK4169.L52   1991   621.32'11--dc20   91-14987
                                                CIP

Published by The Fairmont Press, Inc.
700 Indian Trail
Lilburn, GA 30247

Printed in the United States of America

10 9 8 7 6 5 4 3 2 1

ISBN 0-88173-137-4   FP

ISBN 0-13-529678-1   PH

Distributed by Prentice-Hall, Inc.
A Simon & Schuster Company
Englewood Cliffs, NJ 07632

Prentice-Hall International (UK) Limited, London
Prentice-Hall of Australia Pty. Limited, Sydney
Prentice-Hall Canada Inc., Toronto
Prentice-Hall Hispanoamericana, S.A., Mexico
Prentice-Hall of India Private Limited, New Delhi
Prentice-Hall of Japan, Inc., Tokyo
Simon & Schuster Asia Pte. Ltd., Singapore
Editora Prentice-Hall do Brasil, Ltda., Rio de Janeiro

Contents

Acknowledgements

The information contained in this book has been obtained from a wide variety of authorities who are specialists in their respective fields. Appreciation is expressed to all those who have contributed their expertise to this volume. Many of the chapters in this volume were originally presented at the World Energy Engineering Congress sponsored by the Association of Energy Engineers.

Contributors

Anil Ahuja
G. D. Ander
Jerry D. Andis, C.E.M.
D. Arasteh
S. M. Berman
R. D. Clear
Barbara Coulam
D. L. Dibartolomeo
N. K. Falk
John L. Fetters, C.E.M.
M. S. Gould
C. Greene
D. D. Hollister
A. J. Hunt
R. P. Jarrell
R. L. Johnson
H. Keller
J. J. Kim
William R. King
J. H. Klems
Roger L. Knott, P.E.
C. M. Lampert
D. J. Levy
M. G. Lewis
F. Li
J. L. Lindsey

K. Loffus
M. Milne
O. C. Morse
K. Papamichael
David Peterson
C. P. Quinn
M. D. Rubin
F. M. Rubinstein
M. Schiler
S. E. Selkowitz
M. J. Siminovitch
F. M. Smith
R. E. Snider
M. Spitzglas
R. Sullivan
P. Tewari
R. A. Tucker
W. C. Turner
R. R. Verdeber
Paul von Paumgartten
G. J. Ward
R. E. Webb
G. C. Whalen
R. Whiteman
G. M. Wilde
Jorge B. Wong-Keomt

SECTION I
LIGHTING SYSTEMS DESIGN

Chapter 1
How to Design a Lighting System

With the increased concern for energy conservation in recent years, much attention has been focused on lighting energy consumption and methods for reducing it. Along with this concern for energy efficient lighting has come the realization that lighting has profound affects on worker productivity as well as important aesthetic qualities. This chapter presents an introduction to lighting design and some of the energy efficient techniques which can be utilized while maintaining the quality of illumination.

LAMP TYPES

There are six different light sources that are popular today: incandescent, fluorescent, mercury vapor, metal halide, high pressure sodium and low pressure sodium. All lamps except incandescent are gas discharge lamps, meaning that light is created through the excitation of gases inside the lamp. All gas discharge lamps require a ballast. A ballast accomplishes the following functions:

1. Limits the current flow.
2. Provides a sufficiently high voltage to start the lamp.
3. Provides the correct voltage to allow the arc discharge to stabilize.
4. Provides power-factor correction to offset partially the the coils' inductive reactance.

Lamp efficacy is determined by the amount of light, measured in lumens, produced for each watt of power the lamp requires. The lumens per watt (LPW) of the various light sources can vary considerably. Table 1-1 shows typical LPW ratings including power consumed by the ballast (ballast losses) where applicable.

Table 1-1. General Lighting Lamp/Ballast Characteristics

Type of Lamp	Wattage Range	Initial Lumens Per Watt Including Ballast Losses	Average Rated Life (Hours)
Low Pressure Sodium	18-180	62-150	12,000-18,000
High Pressure Sodium	35-1,000	51-130	7,500-24,000+
Metal Halide	175-1,500	69-115	7,500-20,000
Mercury Vapor			
Standard	40-1,000	24-60	12,000-24,000+
Self-Ballasted	160-1,250	14-30	10,000-20,000
Fluorescent			
Standard	20-215	63-95	9,000-20,000+
Self-Ballasted	8-44	22-50	7,500-18,000
Incandescent	60-1,500	13-24	750-3,500

INCANDESCENT LAMPS

The incandescent lamp is one of the most common light sources and is also the light source with the lowest efficacy (lumens per watt) and shorest life. This lamp is still popular, however, due to the simplicity with which it can be used and the low price of both the lamp and the fixture. Additionally, the lamp does not require a ballast to condition its power supply, light direction and brightness are easily controlled and it produces light of high color quality.

The most common types of incandescent lamps are: the "A" or standard shaped lamp; the "PS" or pear-shaped lamp; the "R" or refelector lamp; the "PAR" or parabolic-aluminized-reflector lamp and the tungsten-halogen (or quartz) lamp.

Light is produced in an incandescent lamp when the coiled tungsten filament is heated to incandescence (white light) by its resistance to a flow of electric current. The life of the lamp and its light output are determined by its filament temperature. The higher the temperature for a given lamp, the greater the efficacy and the shorter the life. The efficacy of incandescent lamps, however, does increase as the lamp *wattage* increase. This makes it possible to save on both energy and fixture costs whenever you can use one higher wattage lamp instead of two lower wattage lamps.

FLUORESCENT LAMPS

The fluorescent lamp is becoming the most common light source. It is easily distinguished by its tubular design—circular, straight or bent in a "U" shape. In operation, an electric arc is produced between two electrodes which can be several feet apart depending on the length of the tube. The ultraviolet light produced by the arc activates a phosphor coating on the inside wall of the tube, causing light to be produced.

Unlike the incandescent lamp, the fluorescent lamp requires a ballast to strike the electric arc in the tube initially and to maintain that arc. Proper ballast selection is important to optimum light output and lamp life.

Lamp sizes range from four watts to 215 watts. The efficacy (lumens per watt) of a lamp increases with lamp length. Reduced wattage fluorescent lamps and ballasts introduced in the last few years use from 10 percent to 20 percent less wattage than conventional fluorescent lamps.

Fluorescent lamps are available in a wide variety of colors but for most application the cool white, warm white and (newly introduced lite white) lamps produce acceptable color and high efficacy. Since fluorescent lamps are linear light sources with relatively low brightness as compared with point sources (incandescent and high intensity discharge lamps), they are suited for indoor application where lighting quality is important and ceiling heights are moderate.

Fluorescent lamp life is rated according to the number of operating hours per start, for example, 20,000 hours at three hours operation per start. The greater number of hours per start, the greater the lamp life. Because fluorescent lamp life ratings have increased, however, the number of times you turn a lamp on or off has become less important. As a general rule, if a space is to be unoccupied for more than a few minutes, the lamps should be turned off.

HIGH INTENSITY DISCHARGE LAMPS

High intensity discharge (HID) is the term used to designate four distinct types of lamps (mercury vapor, metal halide, high pressure sodium and low pressure sodium). Like fluorescent lamps they produce light by establishing an arc between two electrodes; however, in HID lamps the electrodes are only a few inches apart.

HID lamps require a few minutes (one to seven) to come up to full light output. Also, if power to the lamp is lost or turned off, the arc tube must cool before the arc can be restruck and light produced. Up to seven minutes (for mercury vapor lights) may be required.

MERCURY VAPOR LAMPS

The mercury vapor (MV) lamp produces light when electrical current passes through a small amount of mercury vapor. The lamp consists of two glass envelopes: an inner envelope in which the arc is struck, and an outer or protective envelope. The mercury vapor lamp, like the fluorescent lamp, requires a ballast designed for its specific use.

Although, used extensively in the past, mercury vapor lamps are not as popular as other HID sources today due to its relatively low efficacy. However, because of their low cost and long life (16,000 to 24,000 hours), mercury vapor lamps still find some applications.

The color rendering qualities of the mercury vapor lamp are not as good as those of incandescent lamps. A significant portion of the energy radiated is in the ultraviolet region resulting in a "bluish" light in the standard lamp. Through use of phosphor coatings on the inside of the outer envelope, some of the energy is converted to visible light resulting in better color rendition and use in indoor applications.

Mercury vapor lamp sizes range from 40 to 1,000 watts.

METAL HALIDE LAMPS

The metal halide (MH) lamp is very similar in construction to the mercury vapor lamp. The major difference is that the metal halide lamp contains various metal halide additives in addition to the mercury vapor. The efficacy of metal halide lamps is from 1.5 to 2 times that of mercury vapor lamps. The metal halide lamp produces a relatively "white" light, equal or superior to presently available mercury vapor lamps. The main disadvantage of the metal halide lamp is its relatively short life (7,500 to 20,000 hours).

Metal halide lamp sizes range from 175 to 1,500 watts. Ballasts designed specifically for metal halide lamps must be used.

HIGH PRESSURE SODIUM LAMPS

The high pressure sodium (HPS) lamp has the highest efficacy of all lamps normally used indoors. It produces light when electricity passes through a sodium vapor. This lamp also has two envelopes, the inner one being made of a polycrystalline aluminum in which the light-producing arc is struck. The outer envelope is protective, and may be either clear or coated. The light produced by this lamp is a "golden-white" color.

Although the HPS lamp first found its principal use in outdoor lighting, it now is a readily accepted light source indoors in industrial plants. It also is being used in many commercial and institutional applications as well.

HPS lamp size ranges from 35 to 1,000 watts. Ballasts designed specifically for high pressure sodium lamps must be used.

LOW PRESSURE SODIUM LAMPS

The low pressure sodium (LPS) is the most efficient light source presently available, providing up to 183 lumens per watt. The light in this lamp is produced by a U-shaped arc tube containing a sodium vapor. Its use indoors is severly restricted, however, because it has a monochromatic (yellow) light output. Consequently, most colors illuminated by this light source appear as tones of gray.

Low pressure sodium lamps range in size from 18 watts to 180 watts. Ballasts designed specifically for LPS must be used. The primary use of these lamps is street lighting as well as outdoor area and security lighting. Indoor applications such as warehouses are practical where color is not important.

LUMINAIRE EFFICIENCY

In the previous section, it was seen that a lamp produces an amount of light (measured in lumens) which depends on the power consumed and the type of lamp. Equally important to the amount of light produced by a lamp, is the amount of light which is "usable" or provides illumination for the desired task. Luminaires, or lighting fixtures, are used to direct the light to a usable location, dependent on the specific requirements of the area to be lighted. Regardless of the luminaire type, some of the light is directed in non-usable directions, is absorbed by the luminaire itself or is absorbed by the walls, ceiling or floor of the room.

The coefficient of utilization, or CU, is a factor used to determine the efficiency of a fixture in delivering light for a specific application. The coefficient of utilization is determined as a ratio of light output from the luminaire that reaches the workplane to the light output of the lamps alone. Luminaire manufacturers provide CU data in their catalogs which are dependent on room size and shape, fixture mounting height and surface reflectances. Table 1-2 illustrates the form in which a vendor summarized the data used for determining the coefficient of utilization.

To determine the coefficient of utilization, the room cavity ratio, wall reflectance, and effective ceiling cavity reflectance must be known.

Most data assumes a 20% effective floor cavity reflectance. To determine the coefficient of utilization, the following steps are needed:

(a) Estimate wall and ceiling reflectances.

(b) Determine room cavity ratio.

(c) Determine effective ceiling reflectance (pCC).

Step (a)
Typical reflectance values are shown in Table 1-3.

Table 1-2. Vendor Data for 175 Watt Mercury Vapor Lamp—Medium Spread Deflector

Coefficients of Utilization/Effective Floor Cavity Reflectance 20% (pFC)

% REFLECTANCE EFF. CEIL. (pCC)	WALL (pW)	ROOM CAVITY RATIO									
		1	2	3	4	5	6	7	8	9	10
80	50	0.854	0.779	0.711	0.647	0.591	0.539	0.490	0.446	0.407	0.355
	30	0.828	0.739	0.664	0.594	0.533	0.481	0.432	0.388	0.349	0.296
	10	0.805	0.705	0.626	0.552	0.491	0.440	0.392	0.347	0.309	0.258
70	50	0.832	0.761	0.698	0.635	0.578	0.530	0.483	0.438	0.401	0.349
	30	0.808	0.724	0.653	0.585	0.526	0.475	0.426	0.384	0.345	0.295
	10	0.786	0.695	0.618	0.546	0.486	0.434	0.387	0.344	0.308	0.256
50	50	0.788	0.725	0.669	0.610	0.558	0.511	0.466	0.424	0.388	0.339
	30	0.770	0.696	0.632	0.568	0.513	0.464	0.416	0.375	0.338	0.288
	10	0.754	0.670	0.602	0.534	0.478	0.428	0.382	0.339	0.303	0.253
30	50	0.750	0.694	0.642	0.587	0.539	0.495	0.450	0.412	0.377	0.329
	30	0.736	0.671	0.612	0.552	0.499	0.453	0.408	0.367	0.331	0.282
	10	0.722	0.649	0.586	0.523	0.469	0.421	0.375	0.335	0.299	0.249
10	50	0.716	0.665	0.618	0.566	0.521	0.479	0.438	0.399	0.366	0.319
	30	0.704	0.645	0.592	0.536	0.487	0.442	0.399	0.360	0.325	0.276
	10	0.693	0.628	0.571	0.511	0.460	0.413	0.370	0.330	0.294	0.245

Table 1-3. Typical Reflection Factors

COLOR	REFLECTION FACTOR
White and very light tints	.75
Medium blue-green, yellow or gray	.50
Dark gray, medium blue	.30
Dark blue, brown, dark green, and wood finishes	.10

Steps (b) and (c)

Once the wall and ceiling reflectances are estimated it is necessary to analyze the room configuration to determine the effective reflectances. Any room is made up of a series of cavities which have effective reflectances with respect to each other and the work plane. Figure 1-1 indicates the basic cavities.

Figure 1-1. Cavity Configurations

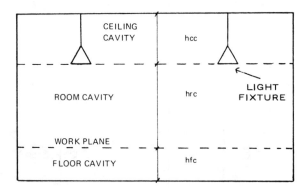

The space between fixture and ceiling is the ceiling cavity. The space between the work plane and the floor is the floor cavity. The space between the fixture and the work plane is the room cavity. To determine the cavity ratio use Figure 1-1 to define the cavity depth and then use Formula 1-1.

(Formula 1-1)   Cavity Ratio = $\dfrac{5 \times d \times (L+W)}{L \times W}$

Where d = depth of the cavity as defined in Figure 1-1
       L = Room (or area) length
       W = Room (or area) width

To determine the effective ceiling or floor cavity reflectance, proceed in the same manner to define the ceiling or floor cavity ratio, then refer to Table 1-4 to find the corresponding effective ceiling or floor cavity reflectance.

SIM 1-1

For Process Plant No. 1, determine the coefficient of utilization for a room which measures 24' X 100'. The ceiling is 20' high and the fixture is mounted 4' from the ceiling. The tasks in the room are performed on work benches 3' above the floor. Use the data in Table 1-2.

Answer

Step (a)
Since no wall or ceiling reflectance data was given, assume a ceiling of .70 and wall of .5.

Step (b)
Assume 3' working height.
hrc = 20-4-3 = 13 (from Figure 1-1)
From Formula 1-1, RCR = 3.4

Step (c)
From Figure 1-1, hcc = 4
From Formula 1-1, CCR = 1
From Table 1-4, pCC = 58
From Table 1-2, Coefficient of Utilization = 0.64
    (interpolated)

Table 1-4. Per Cent Effective Ceiling or Floor Cavity Reflectance for Various Reflectance Combinations

Ceiling or Floor Cavity Ratio	% Ceiling or Floor Reflectance																				
	90				80				70			50			30				10		
% Wall Reflectance	90	70	50	30	80	70	50	30	70	50	30	70	50	30	65	50	30	10	50	30	10
0	90	90	90	90	80	80	80	80	70	70	70	50	50	50	30	30	30	30	10	10	10
0.1	90	89	88	87	79	79	78	78	69	69	68	49	49	48	30	30	29	29	10	10	10
0.2	89	88	86	85	79	78	77	76	68	67	66	49	48	47	30	29	29	28	10	10	10
0.3	89	87	85	83	78	78	75	74	68	66	64	49	48	46	30	29	28	27	11	10	9
0.4	88	86	83	81	78	76	74	72	67	65	63	48	46	45	30	29	27	26	11	10	9
0.5	88	85	81	78	77	75	73	70	66	64	61	48	46	44	29	28	27	25	11	10	9
0.6	88	84	80	76	77	75	71	68	65	62	59	47	45	43	29	28	26	25	11	10	9
0.7	88	83	78	74	76	74	70	66	65	61	58	47	44	42	29	28	26	24	11	10	8
0.8	87	82	77	73	75	73	69	65	64	60	56	47	43	41	29	27	25	23	11	10	8
0.9	87	81	76	71	75	72	68	63	63	59	55	46	43	40	29	27	25	22	11	9	8
1.0	86	80	74	69	74	71	66	61	63	58	53	46	42	39	29	27	24	22	11	9	8
1.1	86	79	73	67	74	71	65	60	62	57	52	46	41	38	29	26	24	21	11	9	8
1.2	86	78	72	65	73	70	64	58	61	56	50	45	41	37	29	26	23	20	12	9	7
1.3	85	78	70	64	73	69	63	57	61	55	49	45	40	36	29	26	23	20	12	9	7
1.4	85	77	69	62	72	68	62	55	60	54	48	45	40	35	28	26	22	19	12	9	7
1.5	85	76	68	61	72	68	61	54	59	53	47	44	39	34	28	25	22	18	12	9	7
1.6	85	75	66	59	71	67	60	53	59	52	45	44	39	33	28	25	21	18	12	9	7
1.7	84	74	65	58	71	66	59	52	58	51	44	44	38	32	28	25	21	17	12	9	7
1.8	84	73	64	56	70	65	58	50	57	50	43	43	37	32	28	25	21	17	12	9	6
1.9	84	73	63	55	70	65	57	49	57	49	42	43	37	31	28	25	20	16	12	9	6
2.0	83	72	62	53	69	64	56	48	56	48	41	43	37	30	28	24	20	16	12	9	6

(more)

6	6	6	6	6	5	5	5	5	5	5	5	5	5	5	5	4	4	4	4	4	4	4	4	4	4	4	4	4	4
9	9	9	9	9	9	9	9	9	8	8	8	8	8	8	8	8	8	8	8	8	8	8	8	8	8	8	8	8	8
13	13	13	13	13	13	13	13	13	13	13	13	13	13	13	13	13	13	13	13	13	13	13	13	14	14	14	14	14	14
16	15	15	14	14	13	13	13	12	12	12	11	11	11	11	10	10	10	10	9	9	9	9	8	8	8	8	8	7	7
20	19	19	19	18	18	18	18	17	17	17	16	16	16	16	15	15	15	15	15	14	14	14	14	14	14	13	13	13	13
24	24	24	24	23	23	23	23	23	22	22	22	22	22	22	21	21	21	21	21	21	20	20	20	20	20	20	19	19	19
28	28	28	28	27	27	27	27	27	27	27	27	27	27	26	26	26	26	26	26	26	26	26	26	25	25	25	25	25	25
29	29	28	27	27	26	26	25	25	24	24	23	23	22	22	21	21	21	20	20	20	19	19	19	19	18	18	18	18	17
36	36	35	35	34	34	33	33	33	32	32	31	31	31	30	30	30	29	29	29	28	28	28	27	27	27	26	26	26	26
43	42	42	42	41	41	41	41	40	40	40	40	39	39	39	39	38	38	38	38	37	37	37	37	37	36	36	36	36	36
40	39	38	37	36	35	34	33	33	32	31	30	30	29	29	28	27	27	26	26	25	25	25	24	24	24	23	23	23	22
47	46	46	45	44	43	43	42	41	40	40	39	39	38	38	37	37	36	36	35	35	34	34	34	33	33	33	32	32	32
56	55	54	54	53	53	52	52	51	51	50	50	49	49	48	48	48	47	47	46	46	46	45	45	45	44	44	44	44	43
47	45	44	43	42	41	40	39	38	38	37	36	35	34	33	33	32	31	30	30	29	29	28	28	27	26	26	25	25	25
55	54	53	52	51	50	49	48	48	47	46	45	44	44	43	42	42	41	40	40	39	39	38	38	37	37	36	36	35	35
63	63	62	61	61	60	60	59	58	58	57	57	56	56	55	54	54	53	53	52	52	51	51	51	50	50	49	49	49	48
69	68	68	67	67	66	66	66	65	65	64	64	64	63	63	62	62	62	61	61	60	60	60	59	59	59	58	58	58	57
52	51	50	48	47	46	45	44	43	42	41	40	39	38	37	36	35	35	34	33	32	32	31	30	30	29	28	28	28	27
61	60	59	58	57	56	55	54	53	52	51	50	49	48	48	47	46	45	45	44	43	43	42	41	41	40	40	39	38	38
71	70	69	68	68	67	66	66	65	64	64	63	62	62	61	60	60	59	59	58	57	57	56	56	55	55	54	54	53	53
83	83	83	82	82	82	82	81	81	81	80	80	80	80	79	79	79	79	78	78	78	78	78	77	77	77	77	76	76	76
2.1	2.2	2.3	2.4	2.5	2.6	2.7	2.8	2.9	3.0	3.1	3.2	3.3	3.4	3.5	3.6	3.7	3.8	3.9	4.0	4.1	4.2	4.3	4.4	4.5	4.6	4.7	4.8	4.9	5.0

Ceiling or Floor Cavity Ratio

LIGHT LOSS FACTOR

The amount of light produced by a luminaire as determined by the lamp lumen output and the fixture coefficient of utilization is the initial value only. Over time the light reaching the task surface will depreciate due to two factors collectively known as the light loss factor (LLF).

The light loss factor (LLF) takes into account that the lumen output of all lamps depreciates with time (LLD) and that the lumen output depreciates due to dirt build-up on the lamp and fixture (LDD). Formula 1-2 illustrates the relationship of these factors.

(Formula 1-2)   LLF = LLF x LLD

To reduce the number of lamps required which in turn reduces energy consumption, it is necessary to increase the overall light loss factor. This is accomplished in several ways. One is to choose a luminaire which minimizes dust build-up. The second is to improve the maintenacne program to replace lamps prior to burn-out. Thus if it is known that a group relamping program will be used at a given percentage of rated life, the appropriate lumen depreciation factor can be found. It may be decided to use a shorter relamping period in order to increase (LLD) even further.

Figure 1-2 illustrates the effect of lumen depreciation and dirt build-up for a typical luminaire. Manufacturer's data should be consulted when estimating LLD and LDD for a luminaire.

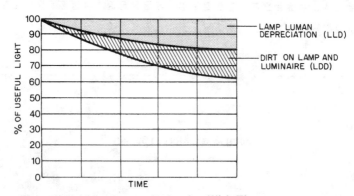

Figure 1-2. Light Output Reduction With Time

ILLUMINATION LEVELS

The amount of light that illuminates a surface is measured in lumens per square foot or footcandles. Table 1-3 shows selected illumination level ranges as recommended by the Illuminating Engineering society in the 1981 Lighting Handbook. Note that these values are recommended for the performance of a specific task and that a room with various task areas would have various recommended illumination levels.

The values in Table 1-3 are intended as guidelines only. The age of the occupants, the inherent difficulty in viewing the object, the importance of speed and/or accuracy for visual performance and the reflectance of the task must be considered when applying these illumination ranges (see IES Handbook for further information).

Table 1-3. Recommended Illumination Values for Selected Areas

Area	Activity	Illuminance Range (Footcandles) on Task
Industrial Assembly	Simple	20-50
	Difficult	100-200
	Extracting	500-1000
Drafting	High Contrast Media	50-100
	Low Contrast Media	100-200
Food Service Facilities	Dining	5-10
Machine Shops	Rough Bench Work	20-50
	Fine Bench Work	200-500
Offices	Lobby	10-20
	Conference Room	20-50
Parking	Open Area	0.5-2
	Closed Area	5-10
Reading	Ditto Copy	50-100
	Ball Point Pen	20-50
	#3 Pencil and Softer Leads	50-100
	Typed Originals	20-50
	Glossy Magazines	20-50

THE LUMEN METHOD

Combining the concepts presented in the previous sections, we can use Formula 1-3 to determine the number of lamps to provide average, uniform lighting levels. This formula is known as the lumen method.

(Formula 1-3) $$N = \frac{E \times A}{Lu \times LLF \times CU}$$

where
  N is the number of lamps required
  E is the required illuminance in footcandles
  A is the area of the room in square feet
  Lu is the lumen output of the lamp
  LLF is the light loss factor which accounts for lamp lumen
    depreciation and lamp (and fixture) dirt depreciation
  CU is the coefficient of utilization

Note that for a specific area and level of illumination for the area, the only means that the lighting designer has for reducing the number of lamps (and consequently the power consumption) required is to use the highest values of Lu, CU and LLF.

SIM 1-2

For the situation described in Example Problem 1, determine the required number of fixtures to give an average, maintained footcandle level of 50. The light loss factor is estimated to be 0.7. The lamps are 175 watt mercury vapor with one lamp per fixture and an initial lumen output of 8,500.

Answer

No. of fixtures $= \dfrac{\text{Area} \times \text{Desired Maintained Footcandle}}{\text{Lumens} \times CU \times LU}$

$= \dfrac{24 \times 100 \times 50}{8,500 \times 0.64 \times 0.7} = 31.5$ or 32 fixtures

THE POINT METHOD

The lumen method is useful in determining the average illumination in an area but sometimes it is desirable to know the illumination level due to one or more lighting fixtures upon a specified point within the area.

The point method (Formula 1-4) computes the level of illumination in footcandles by determining the contribution of a single light source in the area. For multiple light sources, Formula 1-4 must be used for each one and the results summed. Reflections from walls, ceilings and floors are not considered in this method, consequently it is especially useful for very large areas, outdoor lighting and areas where room surfaces are dark or dirty. Additionally, the formula holds true for point sources only. Caution must be exercised when using the point method for fluorescent sources or for luminaires with large reflectors. As a rule of thumb, if the maximum dimensions of the source are no more than one-fifth the distance to the point of calculation, the source will be considered a point source and the calculated illumination will be reasonably accurate.

(Formula 1-4)   Horizontal Footcandles = $\dfrac{cp \times h}{d^3}$
on a Task

where

cp is the candlepower at the desired angle Θ (see Figure 1-3) obtained from manufacturer's data (see Figure 1-4)

h is the height of the fixture above the horizontal plane of the task

d is the distance from the light source to the task

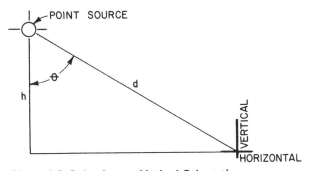

Figure 1-3. Point Source Method Orientations

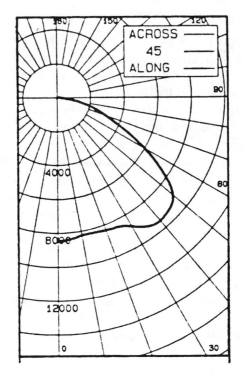

**Figure 1-4. Candlepower Distribution for
A 400 Watt HPS Low Bay Fixture**

A candlepower distribution curve (see Figure 1-4) shows the luminous intensity of a fixture (measured in caldelas) for a range of angular orientations to the fixture. (0° is taken as directly beneath the fixture.)

SIM 1-3

What is the illumination on a surface due to a single 400 watt high pressure sodium light source represented by the data in Figure 1-4 which is 10' in horizontal distance from the workplane? The vertical distance above the workplane (h) is 10'.

Answer

From trigonometric relationships, the angle Θ = arctangent 10/10 = 45°. From Figure 1-4 this gives a candlepower of

9500 candelas. d is found from trigonometric relation-ships to be h/cos Θ = 10/cos 45° = 14.1′. Therefore,

Footcandles = $\dfrac{9500 \times 10}{(14.1)^3}$ = 33.9

FIXTURE LAYOUT

The fixture layout is dependent on the area. The initial layout should have equal spacing between lamps, rows and columns. The end fixture should be located at one-half the distance between fixtures. The maximum distance between fixtures usually should not exceed the mounting height unless the manufacturer specifies otherwise. Figure 1-5 illustrates a typical layout. If the fixture is fluorescent, it may be more practical to run the fixtures together. Since the fixtures are 4 feet or 8 feet long, a continuous wireway will be formed.

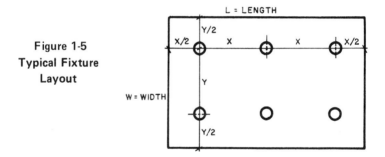

**Figure 1-5
Typical Fixture
Layout**

SIM 1-4

For SIM 1-2 design a lighting layout.

Answer

From SIM 1-2, thirty-two 175 watt mercury vapor lamps are required.

		Rows	Columns	*X* Spacing	*Y* Spacing
Typical Combinations	(a)	4	8	12.5	6
	(b)	3	11	9	8
	(c)	2	16	6	12

(a) 8X = 100   (b) 11X = 100   (c) 16X = 100
       X = 12.5   X = 9   X = 6.2
      4Y = 24   3Y = 24   2Y = 24
       Y = 6   Y = 8   Y = 12

Alternate (b) is recommended even though it requires one more fixture. It results in a good layout, illustrated following.

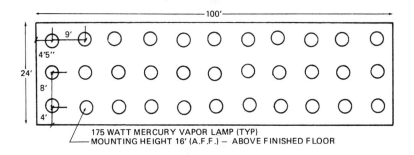

175 WATT MERCURY VAPOR LAMP (TYP)
MOUNTING HEIGHT 16' (A.F.F.) — ABOVE FINISHED FLOOR

CIRCUITING

Number of Lamps Per Circuit

A commonly used circuit loading is 1600 watts per lighting circuit breaker. This load includes fixture voltage and ballast loss. In SIM 1-4, assuming a ballast loss of 25 watts per fixture, a 20 amp circuit breaker, and #12 guage wire, eight lamps could be fed from each circuit breaker. A single-phase circuit panel is illustrated in Figure 1-6. (Note: In practice ballast loss should be based on manufacturer's specifications.)

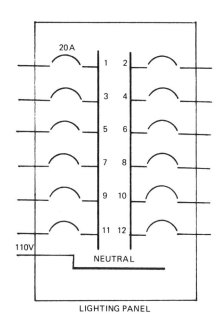

Figure 1-6
Single-Phase
Circuit Panel

LIGHTING PANEL

SIM 1-5

Next to each lamp place the panel designation and circuit number from which each lamp is fed; i.e., A-1, A-2, etc.

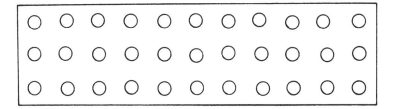

Answer

SIM 1-6

Designate a hot line from the circuit breaker with a small stroke and use a long stroke as a neutral; i.e., ╫╫ 4 wires, 2 hot and 2 neutrals. The lamps are connected with conduit as shown below. Designate the hot and neutrals in each branch.

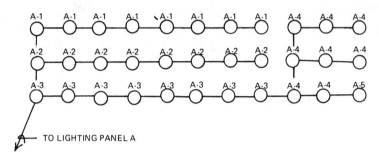

Hint—start wiring from the last fixture in the circuit.

Answer

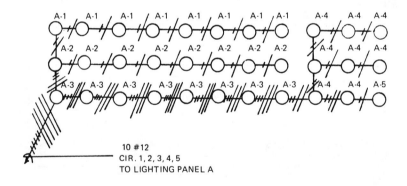

Note that only a single neutral wire is required for each 3 different phase wires.

POINTS ON LIGHTING DRAWINGS

• Choose a lighting drawing scale based on the area to be lighted and the detail required. Typical drawing scale: 1/8″ equals one foot.

- Identify all symbols for lighting fixtures.
- Include circuit numbers on all lights.
- Include a note on fixture mounting height.
- Show "homerun" to lighting panels. "Homerun" indicates the number of wires and conduit size from the last outlet box.
- Use notes to simplify drawing. For example: All wires shall be 2 #12 in 3/4" conduit unless otherwise indicated. Remember the information put on a drawing or specification should be clear to insure proper illumination.

LIGHTING QUALITY

Illumination levels calculated by the lumen and point methods at best give only a "ballpark" estimate of the actual footcandle value to be realized in an installation. Many inaccuracies can be present including: differences between rated lamp lumen output and actual values; difficulty in predicting actual light loss factors; difficulty in predicting room surface reflectances; inaccurate CU information from a manufacturer; non-rectangular shaped rooms.

Precise illumination levels are not critically important, however. Of equal importance to lighting quantity is lighting quality. Very few people can perceive a difference of plus or minus ten footcandles, but poor quality lighting is readily apparent to anyone and greatly affects our ability to comfortably "see" a task.

Of the many factors affecting the quality of a lighting installation, glare has the greatest impact on our ability to comfortably perceive a task. Figure 1-7 shows the two types of glare normally encountered.

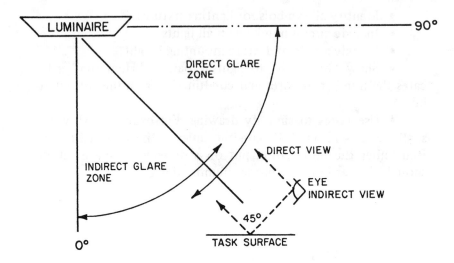

Figure 1-7. Direct and Indirect Glare Zones

DIRECT GLARE

Direct glare is often caused by a light source in the midst of a dark surface. Direct glare also can be caused by light sources, including sunlight, in the worker's line of sight. A rating system has been developed for assessing direct glare called Visual Comfort Probability (VCP). The VCP takes into account fixture brightness at different angles of view, fixture size, room size, fixture mounting height, illumination level, and room surface reflectances.

Most manufacturers publish VCP tables for their fixtures. A VCP value of 70 or higher usually provides acceptable brightness for an office situation. Table 1-5 shows a typical VCP table.

Room Size (in ft.)		Luminaires Lengthwise		Luminaires Crosswise	
		Ceiling Height (in feet)			
W	L	8.5	10.0	8.5	10.0
20 x	20	75	72	73	70
	30	75	72	73	70
	40	75	73	73	71
	60	75	73	72	71
30 x	20	78	75	77	73
	30	78	75	76	73
	40	77	75	75	72
	60	76	74	74	72
	80	76	74	73	72
40 x	20	81	78	80	77
	30	79	77	78	76
	40	78	77	76	75
	60	77	76	75	74
	80	77	75	74	73
	100	76	75	74	73
60 x	30	81	79	80	77
	40	79	78	78	76
	60	78	77	76	74
	80	77	76	75	74
	100	77	76	75	73
100 x	40	81	80	80	78
	60	80	78	78	76
	80	78	77	77	75
	100	78	76	76	74

**Wall Reflectance, 50%
Ceiling Cavity Reflectance, 80%
Floor Cavity Reflectance, 20%
Work Plane Illumination, 100fc

Table 1-5.

INDIRECT GLARE

Indirect glare occurs when light is reflected off of a surface in the work area. When the light bounces off a task surface, details of the task surface become less distinct because contrast between the foreground and background, such as the type on this page and the paper on which it is printed, is reduced. This is most easily visualized if a mirror is placed at the task surface and the image of a light fixture is seen at the normal viewing angle.

This form of indirect glare is called a *veiling reflection* because its effects are similar to those that would result were a thin veil placed between the worker's eyes and the task surface. Veiling reflections can be reduced by:

1) Orient fixtures (or work surface) so that the light produced is not in the indirect glare zone (generally to the side and slightly behind the work position gives the best results).

2) Select fixtures which direct the light above the worst veiling angles (generally 30° or greater). These fixtures have "batwing" distribution patterns such as shown in Figure 1-4. Note, that in selecting fixtures to minimize indirect glare, care must be taken not to select fixtures that are a source of excessive direct glare.

ENERGY CONSERVATION CONSIDERATIONS

Recent concern for energy conservation has focused attention on lighting as an area for potential savings since it can account for 25% of total energy use in an office building. This attention has resulted in many new products which greatly decrease the amount of energy needed for lighting. Unfortunately, blanket application of energy conserving techniques has also resulted in some poorly lit applications which save energy at the expense of worker productivity.

LAMP/LUMINAIRE EFFICIENCIES

As noted in the previous sections, there are wide variations in the efficacies of light sources. By selecting the most efficient light source within the color and room configuration constraints, significant energy savings are possible. Additionally, a new generation of "energy efficient" fluorescent lamps and ballasts are available which offer 5-20% savings over their standard counterparts.

Also, as shown in the previous sections, the choice of luminaire can have a great impact on the energy used for lighting since it determines how much light reaches the task.

NON-UNIFORM LIGHTING

The lumen method presented previously is useful for calculating average uniform illumination for an area, but illumination levels presented in Table 1-3 are for specific tasks in an area. By tailoring illumination levels to the various tasks in an area, significant energy savings are possible.

For example, if 30% of an office area is comprised of desks at which people will be reading material written in pencil, the recommended illumination level is from 50 to 100 footcandles depending on the application. In the remaining 70% of the office, however, if the area is used for general passageways or as a lobby area, the required illumination level is only 10 to 20 footcandles. By directing high levels of illumination only to the task areas, significant energy savings are possible. (Note, that it is generally recommended to limit the ratio of task levels to non-task levels to 3 to 1 to minimize fatigue caused by excessive contrast.)

This technique is referred to as non-uniform or task-ambient lighting. It can be accomplished by positioning fixtures over the task location or by providing a small fixture which is mounted on the desk or machine tool, for example, to provide localized lighting.

GROUP RELAMPING

As seen in Figure 1-2 light output of a fixture depreciates over time due to lamp aging and dirt accumulation. If a minimum footcandle level is desired, initial footcandles must be as much as 40% higher than that desired. Consequently, many more fixtures and consequent power consumption is required to compensate for these depreciation factors.

If a systematic program is initiated, however, to periodically clean the fixtures and relamp before the end of rated life, the number of fixtures can be reduced while maintaining the desired illumination level. This technique is known as group relamping and lighting maintenance. It has the effect of raising the light loss factor (LLF) in Formula 1-3.

LEVEL CONTROLS

Areas with daylight available through windows and skylights can achieve significant energy savings by reducing the lighting system output to maintain a desired illumination level. This can be accomplished by either turning the fixture off, by reducing its output by switching off some of the lamps in the fixture or by using special dimming circuitry.

Additionally, level controls can reduce lighting system energy consumption during "non-production" times when lights are needed. For example, an office which required 70 footcandles during business hours only required 20 footcandles for cleaning at night. By reducing the light levels to 20 footcandles during the cleaning periods, significant energy savings are possible.

Also, level controls (specifically dimming controls) can be used to compensate for light depreciation factors thereby providing required footcandle levels with the minimum possible power consumption.

ON/OFF CONTROLS

One of the simplest and most effective means of controlling lighting energy consumption is by turning off the lights when not needed. To effectively accomplish this may require the addition of switches in each office, grouping of lights into "zones" of usage types, the use of occupancy sensors (either ultrasonic or infrared) to detect when occupants are present and/or the use of an energy management system to automatically schedule lighting operation.

JOB SIMULATION–SUMMARY PROBLEM

SIM 1-7

(a) The Ajax Plant contains a workshop area with an area of 20′ X 18′6″.

For this area compute the number of lamps required, the space between fixtures, and the circuit layout. Use two 40-watt fluorescent lamps per fixture, 2900 lumens per lamp, light loss factor = .7, 110-volt lighting system, 20-watt ballast loss per fixture, and a fixture length of 2′ X 4′. Use luminaire data of Table 1-6, ceiling height 20′ and a desired footcandle level of 40.

Table 1-6. Coefficient of Utilization
20% Effective Floor Cavity Reflectance

Effective Ceiling Cavity Reflectance	80%			50%		
Wall Reflectance	50	30	10	50	30	10
RCR						
10	.33	.26	.22	.31	.26	.22
9	.43	.35	.27	.40	.35	.29
8	.58	.42	.35	.48	.42	.36
7	.58	.50	.42	.55	.48	.42
6	.64	.57	.49	.61	.54	.47
5	.72	.65	.59	.65	.60	.56
4	.77	.71	.64	.71	.65	.60
3	.82	.76	.70	.74	.69	.63
2	.87	.82	.77	.78	.74	.70
1	.91	.87	.83	.81	.78	.75
Spacing not to exceed 1 X Mounting Height						

Analysis

The area of the workshop is $18'6'' \times 20'$.

Assume   hfc = 3

             hcc = 3

Therefore,   hrc = 14

Assume 70% ceiling reflectance

            50% wall reflectance

The room cavity ratio is 7 and the effective ceiling cavity ratio pcc = 53. Thus C.U. = .55.

$$\text{No. of Fixtures} = \frac{20' \times 18\frac{1}{2}' \times 40}{2 \times 2900 \times .55 \times .70} = 7$$

Each 20 amp lighting circuit can provide power for up to 16 fixtures.

Layout Spacing

$3x = 20$
$x = 6.6$
$3y = 18½'$
$y = 6'2''$

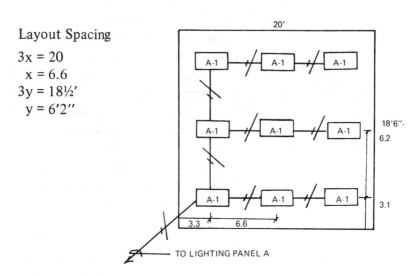

Note: With emphasis on energy conservation, a lighting layout
using 6 fixtures may be preferable.

SUMMARY

Energy conservation is influencing lighting design. Increased emphasis is being placed on minimizing lighting energy
use by using lamps and luminaries which have high lumen outputs and coefficients of utilization. Today's lighting systems
incorporate switching and automatic control devices to make it
easy to turn off lights when they are not required. Lighting
systems need to be analyzed on a first and operating cost basis
to insure that the increasing energy costs are taken into account.

Lighting system design must not only consider the quantity
of illumination but also the quality of illumination. The choice
of a luminaire and its location play an important part in comfortably perceiving a task. An awareness of the importance of
quality lighting can result in a visual environment which is
productive as well as energy efficient.

Chapter 2
Selection Criteria for Lighting Energy Management

Roger L. Knott, P.E.

Today there are many tools available to the designer and facility manager to aid in lighting energy management. Even before the energy concerns became critical in the 1970s, the lighting industry had made substantial progress in improved lamp efficacy* and higher lighting system efficiency.

LIGHT SOURCES

Figure 2-1 indicates the general lamp efficiency ranges for the generic families of lamps most commonly used for both general and supplementary lighting systems. Each of these sources is discussed briefly here. It is important to realize that in the case of fluorescent and high intensity discharge lamps, the figures quoted for "lamp efficacy" are for the lamp only and do not include the associated ballast losses. To obtain the total system efficiency, ballast input watts must be used rather than lamp watts to obtain an overall system lumen per watt figure. This will be discussed in more detail in a later section.

Incandescent lamps have the lowest range of lamp efficacies of the commonly used lamps. This would lead to the accepted conclusion that incandescent lamps should, generally, not be used for large area, general lighting systems where a more effi-

---

*As used in this chapter, this term refers to "luminous efficacy of a source of light" which is defined as the quotient of the total luminous flux emitted divided by the total lamp power input. It is expressed in lumens per watt. See also IES Lighting Handbook, 1984 Reference Volume, The Illuminating Engineering Society of North America, New York, N.Y. 10017.

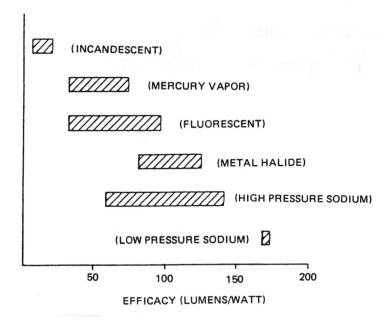

Figure 2-1. General Service Lamp Efficacy

cient source could serve satisfactorily. However, this does not mean that incandescent lamps should never be used. There are many applications where the size, convenience, easy control, color rendering, and relatively low cost of incandescent lamps are suitable for a specific application.

General service incandescent lamps do not have good lumen maintenance throughout their lifetime. This is the result of the tungsten being evaporated off the filament during heating and being deposited on the bulb wall, thus darkening the bulb and reducing the lamp lumen output. Tungsten halogen (quartz) lamps do not suffer from this problem because they use a halogen regenerative cycle so that the tungsten driven off the filament is redeposited on the filament rather than the bulb wall. Therefore, the tungsten-halogen lamps retain lumen outputs in excess of 95 percent of initial values throughout their lifetime.

Mercury vapor lamps find limited use in today's lighting systems because fluorescent and other high intensity discharge (HID) sources have passed them in both lamp efficacy and system efficiency. Typical ratings for mercury vapor lamps range from about 30 to 70 lumens per watt. The primary advantages of mercury lamps are a good range of color, availability in sizes as low as 30 watts, long life and relatively low cost. However, fluorescent systems are available today which can do many of the jobs mercury used to do and they do it more efficiently. There are still places for mercury vapor lamps in lighting system design, but they are becoming fewer as technology advances in fluorescent and higher efficacy HID sources.

Fluorescent lamps have made dramatic advances in the last 10 years. From the introduction of reduced wattage lamps immediately following the Arab oil embargo of the mid 1970s, to the marketing of several styles of low wattage, compact lamps recently, there has been a steady parade of new products. Lamp efficacy now ranges from about 30 lumens per watt to near 90 lumens per watt. The range of colors is more complete than mercury vapor and lamp manufacturers have recently made significant progress in developing fluorescent and metal halide lamps which have much more consistent color rendering properties allowing greater flexibility in mixing these two sources without creating disturbing color mismatches. The recent compact fluorescent lamps open up a whole new market for fluorescent sources. These lamps permit design of much smaller luminaries which can compete with incandescent and mercury vapor in the low cost, square or round fixture market which the incandescent and mercury sources have dominated for so long. While generally good, lumen maintenance throughout the lamp lifetime is a problem for some fluorescent lamp types.

Metal halide lamps fall into a lamp efficacy range of approximately 75-125 lumens per watt. This makes them more energy efficient than mercury vapor but somewhat less so than high pressure sodium. Metal-halide lamps generally have fairly good color rendering qualities. While this lamp displays some very desirable qualities, it also has some distinct drawbacks including relatively short life for an HID lamp, long restrike time to restart after the lamp has been shut off (about 15-20 minutes at 70°F) and a pronounced tendency to shift colors as the lamp ages. In spite of the drawbacks, this source deserves serious consideration and is used very successfully in many applications.

High pressure sodium lamps introduced a new era of extremely high efficacy (60-140 lumens/watt) in a lamp which operates in fixtures having construction very similar to those used for mercury vapor and metal halide. When first introduced, this lamp suffered from ballast problems. These have now been resolved and luminaries employing high quality lamps and ballasts provide very satisfactory service. The 24,000-hour lamp life, good lumen maintenance and high efficacy of these lamps make them ideal sources for industrial and outdoor applications where discrimination of a range of colors is not critical.

The lamp's primary drawback is the rendering of some colors. The lamp produces a high percentage of light in the yellow range of the spectrum. This tends to accentuate colors in the yellow region. Rendering of reds and greens show a pronounced color shift. This can be compensated for in the selection of the finishes for the surrounding areas and, if properly done, the results can be very pleasing. In areas where color selection, matching and discrimination are necessary, high pressure sodium should not be used as the only source of light. It is possible to gain quite satisfactory color rendering by mixing high pressure sodium and metal halide in the proper proportions. Since both sources have relatively high efficacies, there is not a significant loss in energy efficiency by making this compromise.

High pressure sodium has been used quite extensively in outdoor applications for roadway, parking and facade or security lighting. This source will yield a high efficiency system; however, it should be used only with the knowledge that foliage and landscaping colors will be severely distorted where high pressure sodium is the only, or predominant, illuminant. Used as a parking lot source, there may be some difficulty in identification of vehicle colors in the lot. It is necessary for the designer or owner to determine the extent of this problem and what steps might be taken to alleviate it.

Recently lamp manufacturers have introduced high pressure sodium lamps with improved color rendering qualities. However, as with most things in this world, the improvement in color rendering was not gained without cost—the efficacy of the color improved lamps is somewhat lower, approximately 90 lumens per watt.

Low pressure sodium lamps provide the highest efficacy of any of the sources for general lighting with values ranging up to 180 lumens per watt. Low pressure sodium produces an almost pure yellow light with very high efficacy and renders all colors gray except yellow or near yellow. The effect of this is there can be no color discrimination under low pressure sodium lighting and it is suitable for use in a very limited number of applications. It is an acceptable source for warehouse lighting where it is only necessary to read labels but not to choose items by color. This source has application for either indoor or outdoor safety or security lighting, again as long as color rendering is not important.

In addition to these primary sources, there are a number of retrofit lamps which allow use of higher efficacy sources in the sockets of existing fixtures. Therefore, metal halide or high pressure sodium lamps can be retrofitted into mercury vapor fixtures or self ballasted mercury lamps can replace incandescent lamps. These lamps all make some compromises in operating characteristics, life and/or efficacy.

Figure 2-2 presents data on the efficacy of each of the major lamp types in relation to the wattage rating of the lamps. Without exception, the efficacy of the lamp increases as the lamp wattage rating increases.

The lamp efficacies discussed here have been based on the lumen output of a new lamp after 100 hours of operation or the "initial lumens." Like people, not all lamps age in the same way. Some lamp types, such as lightly loaded fluorescent and high pressure sodium as shown in Figures 2-3 and 2-4, hold up well and maintain their lumen output at a relatively high level until they are into, or past, middle age. Others, as represented by heavily loaded fluorescent, mercury vapor and metal halide, decay rapidly during their early years and then coast along at a

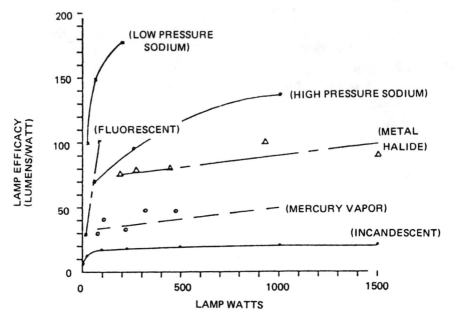

Figure 2-2. Lamp Efficacy
(Does Not Include Ballast Losses)

relatively lower lumen output throughout most of their useful life. These factors must be considered when evaluating the various sources for overall energy efficiency.

LUMINAIRE SELECTION

The luminaire selected can have a significant impact on the energy efficacy of the system as a whole. By selecting a luminaire which generates a distribution of light that results in a high quality and reasonable quantity illuminance, workers will be able to perform their tasks more efficiently. A luminaire which is applied in such a way that direct glare, reflected glare and veiling reflections are kept to a minimum will increase worker productivity.

An important consideration in the efficiency of a lighting system is the Coefficient of Utilization (CU). This is a measure

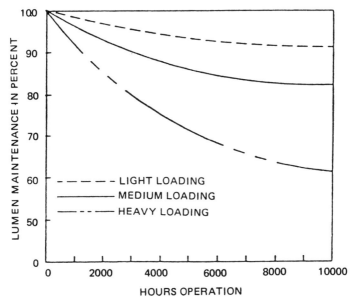

Figure 2-3. Lumen Maintenance Curves
(Typical Fluorescent Lamps)

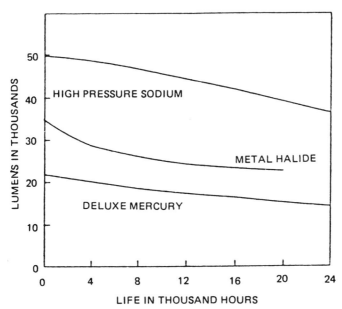

Figure 2-4. Typical Lumen Maintenance Curves
(400 Watt H.I.D. Lamps)

of the efficiency with which the luminaire distributes the lumens generated by the light source to the space. A fixture with a high CU will generally produce higher footcandle levels than one with a lower CU. However, the advantages of a high CU might be lost if the distribution is not carefully controlled. It is often found that luminaires with lower CU's result in lighting systems of superior quality and allow better task visibility and productivity with lower illuminances because of the improved quality. To date there has been no successful method developed which can effectively correlate productivity mathematically with lighting quality or quantity.

With all discharge lamps, fluorescent and HID, it is necessary to employ a ballast to provide starting voltage and control operating current. Advances have been made in recent years which have reduced the ballast losses. Many low loss ballasts are available for fluorescent lamps to provide higher operating efficiencies. Some provide maximum power savings with reduced lamp lumen output and others maintain full light output while reducing input watts.

Some energy efficient systems on the market use three-lamp, 2 ft x 4 ft fluorescent luminaires which provide illuminance equal to four-lamp luminaires employing standard lamps and ballasts. These three-lamp fixtures may reduce the power consumed by 30 percent compared to the standard four-lamp fixtures.

While there have been improvements in HID ballasts over the years, the selection available to the user is, generally, more restricted than in the case of fluorescent.

Finally, the maintainability of the luminaire must be considered as it relates to power and energy efficiency. In industrial facilities particularly, a fixture which is easily maintained is important. Maintenance, generally speaking, is not routinely performed on the lighting system.

If a luminaire is cleaned at all, it generally is cleaned only when it is relamped. With the introduction of longer life lamps, many up to 24,000 hours rated life, relamping may not occur for several years. Without maintenance, the dirt which collects on the luminaire and the lamp can greatly reduce the lumen output of the system.

Some luminaire features should be considered which will improve the maintainability of the system:

Ventilated luminaires which allow, in fact encourage, the passage of air through the lamp chamber help reduce dirt build-up on the luminaire surfaces. This constant flow of air in the bottom and out the top of the reflector tends to carry airborne dirt and contaminants through the fixture without allowing them to settle out on the lamp or reflector. Without this action, oil or grease can collect in greater quantities and allow other dirt particles to stick to the surfaces.

Reflector surfaces can also be effective in reducing the dirt which will adhere. At least one manufacturer realized the importance of a smooth, hard surface many years ago and made the inner surface of their luminaire a smooth glass with the light control prisms on the outer surface. Since the prisms were designed to reflect the light back through the fixture, dirt collecting on the outside, rear surface of the refractor has no impact on the efficiency of the luminaire. Today there are many suitable treatments available, such as Alzak™, Alglas™ and applied silicone surfaces which provide stable, hard, easily cleaned reflector surfaces and contribute to the continued efficiency, both lighting and energy, of the system.

Closures can be provided in particularly dirty environments to stop the dirt and contaminants before they reach the lamp and reflector. Closures can be particularly important in damp, wet or extremely dirty atmospheres where the quantities of contaminant could pose a problem to the continued safe operation of the system, such as situations where steam or water vapor could short out lamp sockets or acid vapors could attack reflector surfaces. Where conditions warrant, the closure should be provided with a gasket.

Breathers can be added to the luminaire to further control the contaminants introduced in enclosed fixtures. Even though a gasketed closure is provided over the bottom of the reflector, as the luminaire alternately heats and cools from being turned on and off, it breathes as the air inside expands and contracts. If this air is drawn through the gasketed perimeter of the closure, it may carry significant quantities of airborne contaminants with it. By adding a breather in the reflector assembly, much of the air passes in and out through the breather and at least a portion of the contaminants will be trapped, thereby reducing the buildup on the luminaire.

SPACE CHARACTERISTICS

Room surfaces can contribute to lighting system power and energy efficiency. This is reasonable considering that much of the illuminance in the space is the result of light interreflected from the room surfaces.

Higher reflectance room surfaces reduce the number of luminaires needed by developing a higher luminaire CU (see Table 2-1). Since the CU affects the Zonal Cavity Calculation for number of luminaires required indirectly, the higher CU recults in fewer fixtures required to provide the desired illuminance as can be seen using the following formula for number of fixtures required:

$$N = \frac{FC \ X \ A}{L/L \ X \ CU \ X \ LLF}$$

Where N = Number of luminaires required
    FC = Illuminance in footcandles (or lux)
    A = Area in square feet (or square meters)
    L/L = Lumens per luminaire
    CU = Coefficient of Utilization
    LLF = Product of all of the Light Loss Factors

Fewer luminaires will, of course, require less power and thereby, consume less energy.

The selection of room surface reflectances requires some care. Low reflectances, as noted above, result in lowered CU and an increased number of luminaires to provide the desired illuminance. However, reflectances which are too high can result in

Table 2-1. Typical Photometrics

COEFFICIENTS OF UTILIZATION　　　ZONAL CAVITY

fc = 20%

cc →	80%				70%				50%			30%		
w \ ρw	70%	50%	30%	10%	70%	50%	30%	10%	50%	30%	10%	50%	30%	10%
1	76	73	71	68	74	72	69	67	69	67	65	66	65	63
2	70	65	61	58	68	64	61	57	62	59	56	60	57	55
3	65	59	54	50	63	58	53	50	56	52	49	54	51	48
4	60	53	48	44	59	52	47	43	50	46	43	49	45	42
5	55	47	42	38	54	47	42	38	45	41	37	44	40	37
6	51	43	37	33	50	42	37	33	40	36	33	40	36	33
7	48	39	33	29	46	38	33	29	37	33	29	36	32	29
8	44	35	29	26	43	35	29	26	34	29	25	33	28	25
9	40	31	26	22	39	31	26	22	30	25	22	30	25	22
10	38	29	23	20	37	28	23	20	28	23	20	27	23	20

glare and/or excessive surface luminance ratios within the space. Glossy finishes should be avoided as they are potential glare sources.

When considering room finishes, such consideration should not be limited to the room surfaces. The equipment located within the space has reflecting surfaces which should be included in considering colors, reflectances and surface finish since they will have an effect upon the lighting system.

Daylighting can be used effectively to reduce electrical lighting energy by the proper use of windows, clerestory, monitor roof or skylights in industrial or office areas. It is usually necessary to provide luminaires throughout the space for necessary lighting during periods when adequate daylight is not available. However, by selective controls, this part of the lighting system can be dimmed or turned off when daylight is available, reducing the energy consumed even though it may not reduce the connected load.

When using daylighting, careful consideration must be given to proper shading and control devices to avoid annoying glare and brightness patterns. It may be possible to increase the effectiveness of daylighting by locating those tasks which can make effective use of daylight in areas where such light is available. Any use of daylighting within the building should be carefully coordinated with the building heating and cooling requirements so that it will be analyzed with the total building energy consumption.

MAINTENANCE

Reference to the formula for the Zonal Cavity method of lighting calculations in the previous section shows there is a term referred to as LLF—Light Loss Factors. Generally speaking, these factors can be grouped into eight categories,[1] as:

1. Luminaire ambient temperature
2. Voltage to luminaire
3. Ballast factor
4. Luminaire surface depreciation
5. Room surface dirt depreciation

6. Lamp lumen depreciation
7. Lamp burnout factor
8. Luminaire dirt depreciation

The first four of these factors can be characterized as "un-recoverable" since there is little that a maintenance program can do to improve them. It may be possible to achieve some improvement in the ambient temperature and voltage which affect the luminaire but it is quite likely that any attempt to significantly affect these factors would not prove to be cost effective. The better solution, if these problems are serious in a given application, would be to use an alternate light source which would be less affected by the particular factor.

Four factors do represent areas which are considered recoverable because they can usually be completely restored by the proper maintenance procedures.

Room surface dirt depreciation (RSDD) is dependent upon the space dirt condition and the procedures used to maintain room surfaces. It was shown previously that the reflectance of the space surfaces has a significant impact on the luminaire CU. The associated surface dirt depreciation factor can also impact the number of luminaires required by the Zonal Cavity Calculation. The conditions affecting the RSDD factor are luminaire distribution type, expected space dirt conditions and the size of the space.

Lamp lumen depreciation (LLD) occurs in all lamps as they age. The lamp lumen depreciation factor is a direct result of the type of lamp and the replacement program followed by the facility. If lamps are spot replaced after they have failed, the LLD factor used for calculating illuminance will be low. However, if there is a group lamping program at the facility, a relamping period can be established to replace lamps at some appropriate time prior to the end of their lifetime. This will result in a higher LLD and can also result in fewer luminaires installed in the space.

Lamp burnout factor (LBO) is directly associated with the maintenance procedures to be followed in the given facility.

This factor does have a relationship with the lamp lumen depreciation (LLD) factor. If the facility chooses to use spot relamping and relamps as each failure occurs, the LLD will be low, because lamps burn to end of life and operate through the most inefficient portion of their life cycle. However, under these conditions, the LBO will be high because lamps will be replaced as they burn out and no sockets will be left with burned-out lamps. On the other hand, if a group relamp method is chosen, lamps will be replaced before their end of life which will raise the average lumens produced per lamp but, since burnouts will not be replaced as they occur, the LBO factor will be lower. The tradeoffs should be considered in reaching a decision on the factors to be used in calculations.

Luminaire dirt depreciation (LDD) is the result of dirt and other airborne comtaminants collecting on the luminaire reflective and enclosing surfaces.

IMPROVEMENT WITH MAINTENANCE

Maximizing each of the listed maintenance factors will improve lighting energy efficiency and cost by:

1. Reducing the initial number of installed fixtures.

2. Reducing the connected power load.

3. Reducing the energy consumed since one of the energy factors, power, has been reduced.

4. Reducing lamp replacement cost with fewer lamps to replace.

5. Reducing luminaire maintenance cost with fewer luminaires to maintain.

Maintenance will be easier with a good design, particularly in industrial areas where there is a reluctance to perform maintenance if it will affect production. Therefore, luminaire location, planned maintenance schedules for lighting and maintenance aids, such as luminaire-lowering devices, can be very effective in reducing maintenance cycles, maintaining maximum preformance of the lighting system and, thereby, contributing to the energy efficiency of the lighting system.

CONTROL OF OPERATIONS

Most of the factors in lighting energy management covered here have been related to the "power" portion of the "energy" equation. The "time" factor relates to the capability to control the lighting system and the hours of operations. Improved energy management may be facilitated if the following measures are considered and implemented where practical:

1. Switch small groups of luminaires to permit operating only those which are necessary for the work being performed.

2. Provide multi-level control of the lighting system within an area by using alternate lamp and/or luminaire switching or multi-level ballasts.

3. Group tasks to place those requiring higher illuminance together and then make use of non-uniform lighting to conserve energy.

4. Use dimming systems to allow variable control of the lighting system for varying tasks.

5. Integrate photocell control devices with switching and/or dimming controls to provide automatic control during periods of adequate daylighting.

6. Install timers to automatically switch lights off during periods when they are not required.

7. Implement power system kilowatt demand controllers to switch off (or down) noncritical lighting loads when the demand sensor approaches the present load limit.

8. Restrict parking to selected areas on second and/or third shifts to reduce the parking lot lighting which must be operated during these hours.

9. Arrange cleaning and maintenance to permit more effective use of lighting energy.

Much has been done in recent years to automate many of these energy conservation techniques. Programmable controllers, demand limiters and automatic timers are commonplace in many large facilities. These systems can be cost effective, particularly when the lighting control forms only a portion of the building systems controlled by these devices. There is a need, however, to perform in-depth cost studies to assure the effectiveness of the substantial initial investment required with some state-of-the-art energy management systems.

A decision cannot be made on the basis of energy consumed by lighting alone—the building heating and cooling costs, plus the productivity of the workers, must be factored into any evaluation. With every energy management system, the return on investment will be improved if the employer and employees are committed to making the system work effectively. Involving the employees in the purposes and operation of the energy management measures will give them a sense of contributing, not only to reducing the company's operating costs but to improving their own quality of life and conserving our vital natural resources.

References
[1] IES Lighting Handbook, 1981 Reference Volume, The Illuminating Engineering Society of North America, New York, N.Y. 10017.

Chapter 3
Fluorescent Lighting —
An Expanding Technology

R.E. Webb, M.G. Lewis, W.C. Turner

INTRODUCTION

Excess light turn-off systems, daylight compensators, infrared motion detectors, No Watt, Phantom Tubes, and power reducers are just a few of the many fluorescent lighting energy saving opportunities available to users today. The potential cost and energy savings offered by many of these lighting products are extremely cost effective. The challenge is to identify which products are applicable to a particular situation and then to effectively evaluate their cost saving potential.

Several years ago lighting manufacturers presented "energy saving" fluorescent bulbs. These bulbs had an increased lighting efficiency, while maintaining the performance characteristics of standard fluorescent bulbs. Many users immediately implemented this opportunity. The cost justification was relatively simple and at the time, it was the only option available short of disconnecting fixtures. Today, users are faced with second- and third-generation "energy saving" fluorescent bulbs, power reducers, and a wide variety of impedence devices. Each successive generation of energy efficient bulbs has boasted increased lighting efficiency, but their applications are more restricted and the cost evaluation is more involved. Therefore, the typical user is faced with more than one alternative which can cost-effectively satisfy their lighting dilemma. Two areas require investigation. First, the user must consider various guidelines for implementation. These guidelines indicate which of the major lighting energy saving opportunities is best applicable, based on esthetics, system flexibility, operating hours and electricity cost. The second area is to investigate the costs of each lighting system.

Another major area receiving increased attention is fluorescent lighting control. As with the new lines of fluorescent bulbs, there are a wide variety of control systems available. These range from infrared motion sensors, designed for private offices to computer-controlled daylight compensating systems which automatically adjust light output to provide a programmed lighting level.

FLUORESCENT LIGHTING SYSTEMS

Lighting energy management involves two interrelated areas. First, the energy analyst must determine if the light levels are correct. Second, the analyst must determine if the most energy efficient, cost effective lighting system is being used. This first section will investigate three basic classifications of fluorescent lighting that may be installed with little or no noticeable reduction in the light output. The second section will then describe lighting systems which increase the lighting efficiency and reduce the light outputs to more acceptable levels.

Energy Efficient Fluorescents

The "energy efficient" fluorescent bulbs have been available for about 10 years. Manufacturers advertise a 15 to 20 percent savings with no noticeable decrease in light levels. The bulbs are available for both 4-foot and 8-foot bulb replacement, with models requiring only 35 watts and 60 watts, respectively. The only possible problem concerns the ballast life. It appears, based on discussions with maintenance managers, if the ballast is old, changing to energy efficient fluorescents may cause the ballast to fail.

Examples of energy efficient bulbs include the SuperSaver by Sylvania, the Watt Mizer by General Electric and the Econo Watt.

Energy Efficient "Plus" Fluorescents

The "energy efficient plus" fluorescents represent the second generation of improved fluorescent lighting. These bulbs are available for replacement of standard 4-foot, 40-watt bulbs and require only 32 watts of electricity to produce essentially the same light levels. To the authors' knowledge, they are not available for 8-foot fluorescent bulb retrofit. The energy efficient plus fluorescents require a ballast change. The light output is similar to the energy efficient bulbs and the two types may be mixed in the same area if desired.

Examples of energy efficient plus tubes include the Super-Saver Plus by Sylvania and General Electric's Watt Mizer Plus.

Energy Efficient Fluorescents-System Change

The third generation of energy efficient fluorescents require both a ballast and a fixture replacement. The standard 2-foot by 4-foot fluorescent fixture, containing four bulbs and two ballasts requires approximately 180 watts (40 watts per tube and 20 watts per ballast). The new generation fluorescent manufacturers claim the following:

- General Electric – "Optimizer" requires only 116 watts with a slight reduction in light output.

- Sylvania – "Octron" requires only 132 watts with little reduction in light level.

- General Electric – "Maximizer" requires 169 watts, but supplies 22 percent more light output.

The fixtures and ballasts designed for the third-generation fluorescents are not interchangeable with earlier generations.

The fluorescent lighting systems described in this second section are designed to reduce the total light output of the system. There are basically four options available.

Option No. 1 – Disconnect Fixtures

The advantages of simply disconnecting lighting fixtures are two-fold—it is very easily done and very cost effective. The disadvantages however, may outweigh the apparent ease of implementation. The lighting plan will appear to be spotty. This may not be a problem in a warehouse area or a corridor, but may not be appropriate for a lobby or open office plan. If this option is selected, remember to disconnect both the bulbs and the ballasts. The ballasts will continue to consume energy even if the bulbs are not connected in the fixture.

Option No. 2 — Power Reducers/No Watt

Power Reducers and No Watts are solid state devices which are connected in series with the ballasts of fluorescent fixtures. The devices are available in 50 of 25 percent reduction models. This means the light level and the energy consumption of the lighting system will be reduced by 50 or 25 percent, respectively. The primary advantage of these devices is they provide evenly reduced lighting. The primary disadvantage is they will not support energy efficient bulbs. Some models must be hardwired into the ballast circuit, while others clip onto the tube connection inside the fixture.

Option No. 3 — Thrift Mates

Thrift Mate is the trade name of a bulb developed by Sylvania. The bulbs come in two types—a 33 percent and a 50 percent reduction model. The bulbs are installed in tandem with standard fluorescent bulbs. The Thrift Mate reduces the energy requirements and light output of both bulbs. The advantages include: the system is very flexible and it produces even lighting. The only potential disadvantage is it appears to have limited use with some ballasts.

Option No. 4 — Phantom Tubes

The phantom tube is a tube which does not light, but simply completes the electrical circuit. It is placed in tandem with a standard bulb and allows only that bulb to light, thus reducing the light output by 50 percent. The advantage of the phantom tube is flexibility, while the disadvantages include uneven lighting and limited use with some ballasts.

The following table illustrates suggested application guidelines.

Table 3-1. Suggested Lighting Guidelines

Category	Description
1	Uneven lighting acceptable, esthetics not a factor.
2	Uneven lighting acceptable, do not disconnect fixtures.
3	Even reduction in lighting required but flexibility not a factor.
4	Even reduction in lighting and flexibility to change required.

Reduction Desired	Category			
	1	2	3	4
15%	D	PT	PR	NW
25%	D	PR	PR	NW
33%	D	PT	PR	TM
50%	D	PT	TM	TM

Key
PT — Phantom Tube
PR— Power Reducer
TM— Thrift Mate
NW— No Watt (Clip-on type)
D — Disconnect

FLUORESCENT LIGHTING CONTROL SYSTEMS

The control of fluorescent lighting systems is receiving increased attention. Two major categories of lighting control are available—personnel sensors and lighting compensators.

Personnel Sensors

There are three classifications of personnel sensors—ultrasonic, infrared and audio.

Ultrasonic sensors generate sound waves outside the human hearing range and monitor the return signals. Ultrasonic sensor systems are generally made up of a main sensor unit with a network of satellite sensors providing coverage throughout the lighted area. Coverage per sensor is dependent upon the sensor type and ranges between 500 and 2,000 square feet. Sensors may be mounted above the ceiling, suspended below the ceiling or mounted on the wall. Energy savings are dependent upon the room size and occupancy. Advertised savings range from 20 to 40 percent.

Several companies manufacture ultrasonic sensors including Novita and Unenco.

Infrared sensor systems consist of a sensor and control unit. Coverage is limited to approximately 200 square feet per sensor. Sensors are mounted on the ceiling and usually directed towards specific work stations. They can be tied into the HVAC control and limit its operation also. Advertised savings range between 30 and 50 percent.

Audio sensors monitor sound within a working area. The coverage of the sensor is dependent upon the room shape and the mounting height. Some models advertise coverage of up to 1,600 square feet. The first cost of the audio sensors is approximately one-half that of the ultrasonic sensors. Advertised energy savings are approximately the same as the ultrasonic sensors. Several restrictions apply to the use of the audio sensors. First, normal background noise must be less than 60 dB. Second, the building should be at least 100 feet from the street and may not have a metal roof.

Lighting Compensators

Lighting compensators are divided into two major groups—switched and sensored.

Switched compensators control the light level using a manually operated wall switch. These particular systems are used frequently in residential settings and are commonly known as "dimmer switches." Based on discussions with manufacturers, the switched controls are available for the 40 watt standard fluorescent bulbs only. The estimated savings are difficult to determine, as usually switched control systems are used to control room mood. The only restriction to their use is the luminaire should have a dimming ballast.

Sensored compensators are available in three types. They may be very simple or very complex. They may be integrated with the building's energy management system or installed as a stand alone system. The first type of system is the Excess Light Turn-Off (ELTO) system. This system senses daylight levels and automatically turns off lights as the sensed light level approaches a programmed upper limit. Advertised paybacks for these types of systems range from 1.8 to 3.8 years.

The second type of system is the Daylight Compensator (DAC) system. This system senses daylight light levels and automatically dims lights to achieve a programmed room light level. Advertised savings ranges from 40 to 50 percent. The primary advantage of this system is it maintains a uniform light level across the controlled system area.

The third system type is the Daylight Compensator + Excess Light Turn-Off system. As implied by the name, this system is a combination of the first two systems. It automatically dims light outputs to achieve a designated light level and, as necessary, automatically turns off lights to maintain the desired room conditions.

SUMMARY

Fluorescent lighting is the most widely used commercial lighting source in the United States. The technology of fluorescent lighting is advancing rapidly and may have already made much of this chapter obsolete. Energy managers must continue to review the newest types of bulbs and controls to insure their buildings are operating as efficiently as possible.

Chapter 4
Practical Selection of Fluorescent Lamps with Emphasis on Efficiency and Color

R.E. Snider

During the last few years numerous new fluorescent lamps have been introduced. Today, there is a wide selection of lamps of different wattages and colors available. The efficiency of these lamps can vary significantly. The purpose of this chapter is to investigate the choices which are available and give guidance in the practical selection of fluorescent lamps. Since most of the fluorescent lamps used today are 4' in length versus 8', this chapter will deal mainly with 4' lamps.

The concepts presented in this chapter are that of the author. Generalizations will often be made to present ideas in a practical manner. The author uses utility rates and construction data for Alabama. Each analysis should use the cost data for the specific location in question.

Although 4' lamps are available in other wattages we will be comparing 40W, 34W, and 32W lamps as these are the most popular. We will first compare cool white lamps since they comprise the majority of lamps sold. When comparing lumen output of lamps of the same wattage from three different manufacturers, they generally differed by 0-7%. Most lumen outputs differed no more than 3%. Since design is usually done with worst case situations, the minimum lumen output rather than the average lumen output was selected for comparison purposes. Prices for lamps, fixtures, labor rates, energy costs, etc. are approximate prices and are given only to allow comparisons to be made in a realistic manner. Prices, wage rates and labor costs may vary drastically from region to region and these variables should be adjusted for each geographic area.

In the economic analysis we will generally consider the number of fixtures, the installed cost of the fixture with lamps, replacing lamps every 5 years, and energy costs. Economic decisions will be based on the present worth method of comparison.

We will first compare 40W lamps versus 34W lamps. Comparisons will be made using economy and premium grade troffers with energy saving ballasts. Since the 40W lamp produces 15% more light than the 34W lamp, it will take 15% more 34W lamps to produce the same amount of light; thus 15% more fixtures. We will assume a cost of 4¢/kW hour. We will choose 3 lamp troffers since this is a good average between 2, 3, and 4 lamp troffers. We are concerned with the final cost to the owner and therefore will be using cost after the electrical contractor and general contractor have added overhead and profit. We will consider a combined mark-up of 30%. We will assume it takes 1 manhour to install one fixture at a labor rate of $9.50/hour. The average life of this type of lamp is 20,000 hours. Assuming 12 hours/day and 6 day/week operation, the lamp life is 5.3 years. We will assume a 5-year life. An interest rate of 8% compounded annually will be used.

	Economy Grade Troffer		Premium Grade Troffer	
	34W	40W	34W	40W
Initial Cost	+20%	–	+19%	–
5-Year Cost	+7%	–	+8%	–
10-Year Cost	+5%	–	+6%	–
15-Year Cost	+5%	–	+6%	–

Therefore, based on the assumptions we have made, 34W lamps are not economical for this area. Where energy costs are considerably higher they will be more economical.

Next we will compare the 40W/CW lamp to the standard size 32W/T-12/CW lamp. For simplicity only premium grade troffers will be used. If premium grade fixtures prove economical, then economy grade fixtures will also be economical.

It should be noted that 32W/T-12 lamps operate a little differently from standard 34W and 40W lamps. If the lamps are switched on from a cold start and remain on, the operation is the same. However, if the lamps are switched off and then switched back on within one minute, it may take 1-5 minutes for the lamps to restart and come back to full brightness. This could be critical if the lamps are on an emergency circuit connected to a generator and expected to be back on in no less than 10 seconds. It also causes nuisance problems by people who are not familiar with their operation. For example, if they turn the lights off, remember they have forgotten something and turn the light switch back on, the lamps will be very dim and have uneven light output. They may think something is wrong and call maintenance.

	40W	32W
Initial Cost	–	+24%
5-Year Cost	–	+9%
10-Year Cost	–	+7%

Here again 40W lamps appear the most economical for this area.

Next we will compare the standard 40W/CW to the 32W/-CW/T-8. The T-8 lamp is similar to the standard 40W/T-12 except the T-8 is 1″ in diameter rather than 1½″. Since there is less lamp to block the reflected light, the fixture is therefore more efficient. Coefficients of utilization will be higher for T-8 lamps than T-12 lamps. The fixture is about 9% more efficient with T-8 lamps versis T-12 lamps. Because of this a 40W/CW can virtually replace a 32W/T-8 lamp. The T-8 lamps require a different ballast than the standard lamps. T-8 lamps will not work properly on ballasts for standard lamps and vice versa. Also 4-foot T-8 lamps are only available in 32W. The 40W/T-8 lamps are 5 feet in length. You can expect an adder to the fixture cost where T-8 lamp ballasts are required. We will first consider magnetic ballasts and assume an adder of $7.50 per ballast (contractor cost) over energy saving ballasts.

	40W	32W/T-8
Initial Cost	–	+30%
5-Year Cost	–	+5%
10-Year Cost	–	+0%
15-Year Cost	–	+1%

Again the standard 40W lamp appears the most economical choice for this area. The increase in cost between 10-15 years is due to the fact that the ballast needs to be replaced at the end of 10 years.

Next we will consider electronic ballasts to see if this effects lamp choice. Some electronic ballast manufacturers make ballasts which will operate 1 or 2 lamps while others make an electronic ballast which will operate 1, 2, 3, or 4 lamps.

We will first compare a fixture with standard 40W/CW lamps and 2 energy saving ballasts versus a fixture with standard 40W/CW lamps and 2 electronic ballasts. We will assume an adder of $22 per ballast (contractor cost) over energy saving ballasts.

	40W/ESB	40W/Electronic(2)
Initial Cost	–	+66%
5-Year Cost	–	+22%
10-Year Cost	–	+11%
15-Year Cost	–	+1%

40W lamps with energy saving ballasts were again more economical.

We will not compare a fixture with standard 40W/CW lamps and 2 energy saving ballasts versus a fixture with standard 40W/CW lamps and one 3-lamp electronic ballast. We will assume an adder of $24 (contractor cost) for the electronic ballast over the fixture with energy saving ballast.

	40W/ESB	40W/Electronic(1)
Initial Cost	—	+36%
5-Year Cost	—	+4%
10-Year Cost	+5%	—
15-Year Cost	+14%	—

If you can afford the 36% increase in initial cost, the single electronic ballast should start saving money in 7-8 years.

34-watt lamps are not recommended for use on some electronic ballasts because of starting problems and therefore will not be considered.

Next we will compare the standard 40W/CW lamp versus the 32W/CW/T-8 lamp with 1 electronic ballast. The single electronic ballast costs only a little more for the T-8 lamps versus the T-12 lamps. We will assume an adder of $26 (contractor cost) over energy saving ballast.

	40W	32W/T-8/Electronic(1)
Initial Cost	—	+47%
5-Year Cost	—	+6%
10-Year Cost	+3%	—
15-Year Cost	+12%	—

Again the initial cost of the electronic ballast is extremely high, but after 8-9 years it should start paying off. Also note that the standard 40W lamps with electronic ballasts pay off sooner with a lower first cost than the 32W/T-8 lamps.

Note: A lamp life of 20,000 hours using electronic ballasts has been assumed. Lamp life may need to be reduced to 15,000 hours depending on the lamp and ballast combination used.

Up to now we have assumed energy saving ballasts to be more cost effective than standard ballasts. Let us verify this assumption. We will assume a top-of-the-line energy saving ballast with an adder per ballast of $2.50.

	Economy Grade Troffer		Premium Grade Troffer	
	40W/Standard	40W/E.S.B.	40W/Standard	40W/E.S.B.
Initial Cost	—	+12%	—	+8%
5-Year Cost	+3%	—	+3%	—
10-Year Cost	+6%	—	+5%	—

Therefore, the energy saving ballast costs more initially but pays off in less than 5 years and is a good choice.

Generally we have found that energy saving products are not always the most economical. This is because more fixtures will usually be required with the energy saving lamps. The cost of the lamps is also usually higher. When using 32W/T-12 lamps remember they have different warm start characteristics than standard lamps. 32W/T-8 lamps also require special ballasts and are not compatible with standard ballasts. Some fixtures will have higher coefficients of utilizations when using T-8 lamps versus T-12 lamps. 34W lamps are not recommended to be used with some electronic ballasts and these lamps may have lower lamp life.

SUMMARY OF FIXTURES WITH DIFFERENT TYPE CW LAMPS

	40W	34W	32W	32W/T-8	40W/Elect.(2)	40W/Elect.(1)	32W/T-8 Elect.(1)
Initial Cost	—	+19%	+24%	+30%	+66%	+36%	+47%
5-Year Cost	—	+8%	+9%	+5%	+22%	+4%	+6%
10-Year Cost	—	+6%	+7%	+0%	+11%	−5%	−3%
15-Year Cost	—	+6%	−2%	+1%	+1%	−14%	−12%
Watts Used	136	115	109	108	107	97	88
No. of Fixtures	1.0	1.15	1.17	1.0	1.0	1.0	1.0

NOTE: *The above data is based on utility rates and construction costs for Alabama. Each job will require a cost analysis based on the specific application and locality. Figures may differ significantly from those shown above.*

COLOR CONSIDERATIONS

The color of fluorescent lamps is generally described by the correlated color temperature expressed in Kelvin degrees and the color rendering index of C.R.I. expressed as a whole number from 1-100.

The International Commission on Illumination (CIE) developed a diagram called the CIE Chromaticity Diagram to aid in specifying colors. Plotted on the diagram is a curved line called the Black Body Locus. This line represents the change in color of a piece of metal as it is heated. Therefore, a particular color can be specified by indicating the degree Kelvin provided it is an incandescent lamp.

The spectrum from fluorescent lamps does not fall exactly on the curve; therefore, the correlated color temperature must also be specified. The problem with this method is that one lamp could plot above the curve where another plots below the curve. Both could have the same correlated color temperature while having different color characteristics. Therefore, temperature is not an exact measure of color for fluorescent lamps.

To aid in specifying colors, a system called the color rendering index (CRI) was developed. The purpose of the color rendering index is to measure the shift from the curve on the CIE chromaticity diagram. This is done by measuring the shift of eight specified colors compared to a reference source of the same degree Kelvin. The shift in each color is determined and then the numbers are averaged to determine the CRI. Therefore, one lamp could score high in reds and another high in blues and yet have the same average score. This is not common but it can happen.

Though cumbersome, sometimes the best method to specify fluorescent lamps is to examine the color spectrum of a lamp by its spectral power distribution curves. From these curves it is easy to identify that cool white lamps have more blue and less red than warm white lamps. Also, you can see that deluxe cool white lamps are much stronger in reds than cool white or warm white. These lamps have a more even distribution of colors but are not as good as the 5000°K lamps.

From the spectral distribution curves you can see the difference in the single coated lamps and the fairly new double coated tri-phosphor lamps. The double coated lamps have one coat of the same phosphors as the regular lamps and then they have a second coat of more expensive phosphors. The second coat of phosphors contains three strong primary colors producing phosphors in the yellow-red, green, and blue-green wave lengths. The manufacturer's design philosophy is that most colors can be made by a combination of these three colors and therefore the color rendering ability of the lamp is improved. These lamps provide an economical way of improving the CRI while maintaining high lumen output. These are available at 3000°K, 3500°K, and 4100°K.

In choosing a fluorescent lamp there is no right or wrong color. We would like a fluorescent lamp which renders colors as we are used to viewing them. For example,, at home we are normally viewing objects from a 3000°K source; at work from a 4000°K source and outside the daylight varies from 1800°K to 28,000°K.

A lot of objects simply do not appear as pleasing when viewed under natural daylight as they do under other light sources. Therefore selection of lamps is a matter of personal preference in most cases.

In the following section, 10 lamp color comparisons will be made. Comparisons will include temperature, CRI, lumens, lamp cost and 5-year cost for the following:

1. CW vs WW
2. CW vs LW
3. CW vs 5000°K
4. WW vs LW
5. WW vs 5000°K

6. CWX vs 5000°K
7. CW vs SP35
8. WW vs. SP35
9. CW vs SP41
10. WW vs SP30

No. 1: CW vs WW

	Watts	Approx. Temp.	Approx. CRI	Approx. Lumens	Approx. User Lamp Cost	No. of Fixtures	Approx. 5-Year Cost
CW	40	4200	62	3150	2.50	+2%	+1%
WW	40	3000	52	3200	3.30	—	—

Remarks: Clothes and skin will generally look better under WW. WW gives skin a slight tan and brings out the reds in clothes.

No. 2: CW vs LW

	Watts	Approx. Temp.	Approx. CRI	Approx. Lumens	Approx. Lamp Cost	No. of Fixtures	Approx. 5-Year Cost
CW	40	4200	62	3150	2.50	—	—
LW	34	4200	49	2925	3.75	+8%	+2%

Remarks: CW definitely has better color. LW is not available in 40W. The LW has a high lumen per watt output but still does not prove economical according to the figures used.

No. 3: CW vs 5000°K

	Watts	Approx. Temp.	Approx. CRI	Approx. Lumens	Approx. Lamp Cost	No. of Fixtures	Approx. 5-Year Cost
CW	40	4200	62	3150	2.50	—	—
5000°K	40	5000	90	2000	6.20	+43%	+52%

Remarks: The 5000°K has excellent color rendering characteristics throughout the spectrum. The 5000°K lamp is generally
used where color determination is critical.

No. 4: WW vs LW

	Watts	Approx. Temp.	Approx. CRI	Approx. Lumens	Approx. Lamp Cost	No. of Fixtures	Approx. 5-Year Cost
WW	40	3000	52	3200	3.30	—	—
LW	34	4200	49	2925	3.75	+9%	+10%

Remarks: Although the CRI's are close, there is a big difference
in the lamps' rendering. The LW appears bland while the WW is
rich in red.

No. 5: WW vs 5000°K

	Watts	Approx. Temp	Approx. CRI	Approx. Lumens	Approx. Lamp Cost	No. of Fixtures	Approx. 5-Year Cost
WW	40	3000	52	3200	3.30	—	—
5000°K	40	5000	90	2200	6.20	+45%	+52%

Remarks: The WW may be considered more pleasing while the
5000°K is considered more natural.

No. 6: CWX vs 5000°K

	Watts	Approx. Temp.	Approx. CRI	Approx. Lumens	Approx. Lamp Cost	No. of Fixtures	Approx. 5-Year Cost
CWX	40	4100	89	2100	4.50	+5%	+2%
5000°K	40	5000	90	2200	6.20	—	—

Remarks: Both have excellent color rendering characteristics,
and yield fairly "natural" colors. They do not over-emphasize
any one color. The 5000°K has a more even color balance and
should be used where color is critical such as in nurseries in
hospitals. CWX should be good enough for most exam rooms
and treatment rooms in hospitals.

No. 7: CW vs SP35

	Watts	Approx. Temp.	Approx. CRI	Approx. Lumens	Approx. Lamp Cost	No. of Fixtures	Approx. 5-Year Cost
CW	40	4200	62	3150	2.50	+1%	—
SP35	40	3500	69	3180	4.10	—	+2%

Remarks: The SP35 color rendering is usually more pleasing than the CW. The SP35 offers a good compromise between the red color of WW lamps and the blue color of CW lamps. The SP35 does a good job on clothes and skin. The popularity of this lamp is just beginning to catch on and as demand increases the price will probably drop. The SP35 may be the lamp of choice for general use.

No. 8: WW vs SP35

	Watts	Approx. Temp.	Approx. CRI	Approx. Lumens	Approx. Lamp Cost	No. of Fixtures	Approx. 5-Year Cost
WW	40	3000	52	3200	3.30	—	—
SP35	40	3500	69	3180	4.10	+1%	+2%

Remarks: Again the SP35 is a good compromise between the WW and CW.

No. 9: CW vs SP41

	Watts	Approx. Temp.	Approx. CRI	Approx. Lumens	Approx. Lamp Cost	No. of Fixtures	Approx. 5-Year Cost
CW	40	4200	62	3150	2.50	+3%	+1%
SP41	40	4100	69	3240	3.75	—	—

Remarks: The SP41 offers some improvement in color over the CW. This may be a good lamp if you are concerned that the lamps will be mixed with CW lamps when replacement starts. The lamps will look the same in the fixture. This may also be a good lamp for patient bedrooms. Each bedroom has a window and this °K is a compromise between incandescent and daylight.

No. 10: WW vs SP30

	Watts	Approx. Temp.	Approx. CRI	Approx. Lumens	Approx. Lamp Cost	No. of Fixtures	Approx. 5-Year Cost
WW	40	3000	52	3200	3.30	+1%	
SP30	40	3000	69	3230	4.10	—	

Remarks: The SP30 has a significantly improved CRI and is an excellent choice if you want a 3000°K lamp. Great for clothes and skin color enhancement.

The type of a lamp should not be chosen to highlight the color of wallpaper or carpet in a room. More important is the function of the space and what will be in the room. The designer must consider the overall effect of the space when selecting a lamp. There is no right or wrong choice. It is simply a matter of opinion. The author feels that the SP35 lamps are probably the best choice for general offices and variety retail stores and the SP30 lamps for clothing stores.

The prices used in the calculations for this chapter are based on the approximate cost of materials, labor, and energy in Alabama. These factors will vary significantly from area to area and must be adjusted accordingly. Hopefully this chapter has illustrated the various options available when selecting a fluorescent lamp.

SECTION II
LIGHTING CONTROLS

Chapter 5
Review of Lighting Control[1] Equipment and Applications

R.R. Verderber

Many types of lighting control equipment permit the automatic management of lighting systems. Lighting control equipment can be used to lower light levels, turn lights on and off on a schedule, and respond to the availability of natural light or the presence of occupants. Each of these operations minimizes the energy consumed while providing the proper illumination for use of the space.

The choice of a lighting control system depends on its particular application (retrofit, renovation, or new construction). This chapter will present a basis for selecting the control equipment and the control strategies (light reduction, scheduling, tuning, lumen depreciation, daylighting, load-shedding) most appropriate to an application, along with the important cost factors (initial, operating, installation, supply circuit layout, building design) that must be considered for each application.

LIGHT CONTROL EQUIPMENT

Table 5-1 lists various types of equipment that can be components of a lighting control system, with a description of the predominant characteristic of each type of equipment. Static equipment can alter light levels semipermanently. Dynamic equipment can alter light levels automatically over short intervals to correspond to the activities in a space. Different sets of components can be used to form various lighting control systems in order to accomplish different combinations of control strategies.

---

[1]This work was supported by the Assistant Secretary for Conservation and Renewable Energy, Office of Building Energy Research and Development, Buildings Equipment Division of the U.S. Department of Energy under Contract No. DE-AC03-76SF00098.

Table 5-1. Lighting Control Equipment

System	Remarks
STATIC:	
Delamping	Method for reducing light level 50%.
Impedance Monitors	Method for reducing light level 30, 50%.
DYNAMIC:	
Light Controllers	
Switches/Relays	Method for on-off switching of large banks of lamps.
Voltage/Phase Control	Method for controlling light level continuously 100 to 50%.
Solid-State Dimming Ballasts	Ballasts that operate fluorescent lamps efficiently and can dim them continuously (100 to 10%) with low voltage.
SENSORS:	
Clocks	System to regulate the illumination distribution as a function of time.
Personnel	Sensor that detects whether a space is occupied by sensing the motion of an occupant.
Photocell	Sensor that measures the illumination level of a designated area.
COMMUNICATION:	
Computer/Microprocessor	Method for automatically communicating instructions and/or input from sensors to commands to the light controllers.
Power-Line Carrier	Method for carrying information over existing power lines rather than dedicated hard-wired communication lines.

Static Controls. These controls can decrease light levels by a discrete amount, usually 30 or 50 percent. They provide a rapid, economical way to reduce energy consumption in an over-illuminated area. The method can be as simple as removing two lamps from a four-lamp fixture (delamping) or installing impedance-modifying lamps or devices. Some impedance-modifying devices must be placed in the fixture and hard-wired to the supply power, adding to installation costs.

Dynamic Controls. Automatic dynamic lighting control systems can consist of a combination of lighting devices such as controllers, which dim or switch lamps on and off; sensors, hich measure light levels or sense the presence of occupants in a space; and communications, which process information from sensors and pass instruction to the light controllers.

LIGHT CONTROLLERS

Fluorescent lamps can be switched on and off via switches or relays. Because switches or relays must be hard-wired into power lines, they are most economical when they control one phase of a circuit or a large bank of lamps. Some devices can vary the amplitude of line voltage or duty cycle to standard core-coil ballasted fluorescent lamps. The light output of the lamps can then vary continuously from full to about 50 percent. Since they operate by conditioning the supply power, they are best suited to large banks of lamps. The systems are relatively economical to install, since they involve a central control. However, their flexibility is limited by the supply circuit layout.

Solid-state dimming ballasts are designed to continuously control the light output of fluorescent lamps down to 10 percent of full light output. The command to the ballast is via low-voltage signals (0 to 12 volts). Groups of lamps can connect on the same communication link with low-voltage wire. Thus, the flexibility of this method is not limited by the supply circuit layout. Each ballast has a manual adjustment so that maximum light output of each fixture can be achieved. This permits the use of all the lighting control strategies.

Some Control Devices

The sensors in a control system are used to obtain information about occupancy, time, or available daylight. Some sensors can relay commands directly to lighting controllers to operate lamps in a prescribed manner or they can pass information to a control processor. Personnel sensors determine the occupancy of a given space by detecting motion. They are most effective for spaces that have only one occupant, since the greater the number of occupants, the less chance that the space will be unoccupied.

Personnel sensors on the market are self-contained systems that include a sensor and a light controller. Because they must be hard-wired into the supply power, installation costs are involved. They are most effective when used by occupants who spend large portions of their time away from their stations or at stations that are intermittently used by several occupants. The controlled area must be occupied infrequently and unpredictably for personnel sensors to be most effective.

Photocells are used to measure the illumination levels in a space. If there is a change in the prescribed illumination level, i.e., because of lumen depreciation or daylighting, photocells signal the electric lights to maintain the prescribed level. The photocell outputs can be sent to a lighting controller that can alter the light levels or to a central processor. The photocell requires low-voltage wiring that creates some installation costs. Photocell lighting control systems that need to be hard-wired to the supply power incur additional installation costs.

Clocks provide instructions to a lighting system in real time. The equipment can be as simple as a spring-loaded switch that will turn the lights off at a prescribed time. It can also be incorporated in a control processor that will control the lighting system according to a prescribed daily routine for an entire year.

Communication

A single, central processor can store information and satisfy the communication needs for very large areas. The relatively high cost of a unit means it is generally not cost-effective for controlling a small amount of floor space. The computer/microprocessor is centrally located and involves relatively small installation costs.

A power-line carrier is a technically feasible device for transmitting input and output signals to and from a central processor. A carrier is particularly attractive in retrofit applications because it does not require distribution of signal wire. For other applications one must consider the relative costs and advantages of using a power-line carrier or distributing low-voltage signal wire.

LIGHTING CONTROL
EQUIPMENT AND STRATEGIES

Table 5-2 lists the lighting control strategies that can make use of the various types of control equipment with notes on the strategies listed for some equipment. The number of strategies in which a control is most effective depends on the application (Table 5-3). The following points out some of the features of the control equipment.

Static Controls. The static controls, delamping and impedance monitors, can be used to reduce the light levels throughout a space or in selected areas, i.e., by tuning. Since these strategies are most effective in spaces that are over-illuminated, they are best used in retrofit applications.

Dynamic Controls. The dynamic lighting controls include the use of various control components to create a system that executes the types of control strategies that are apropos depending on the application.

Light Controllers. In retrofit applications, relay-type controls are best limited to a single strategy, scheduling. A central processor with a clock can control an entire floor or building. If the power supply distribution can be altered, as in renovations, two control strategies can be accomplished. For appropriate buildings designed to exploit natural light, relay-type controls can employ three strategies (Table 5-2).

Voltage/Phase Control. These devices condition the input power to standard core-coil ballasts, which permits the dimming of fluorescent lamps over a continuous range. Because these devices dim lamps over a continuous range, these systems can accomplish two strategies in retrofit applications (see Table 5-2). If photocells are used, three startegies can be accomplished, including lumen depreciation.

Table 5-2. Lighting Control Equipment

System	Application		Strategy						
		Light Reduction	Scheduling			Lumen Deprec.	Day-lighting	Load-Shedding
			Predict.	Random	Tuning			
STATIC								
Delamp	Retrofit	X			X			
Impedance Modifier	Retrofit	X			X			
DYNAMIC								
Light Controller								
Switch/Relay	Retrofit		X					
	Renovation		X					X
	New Construction		X					X
Voltage/Phase Control	Retrofit	X	X					
	New Construction		X				X	X
Solid-State	Renovation				X	X	X	X
Dim. Ballast	New Construction				X	X	X	X
Sensors								
Clocks	Retrofit		X					
	Ren./New Const.		X					
Personnel	Retrofit			X				
	Ren./New Const.			X				
Photocell	Retrofit					X		
	Ren./New Const.					X	X	
Communication								
Computer/ Microprocessor	Retrofit		X			X		
	Renovation		X			X	X	X
	New Construction		X				X	X
Power-Line Carrier	Retrofit		X			X	X	
	Ren./New Constr.		X		X			X

Table 5-3. Major Cost Factors for Lighting Control Applications

Application	Initial Cost	Operating Cost	Installation Cost	Supply Circuit Layout	Building Design
Retrofit	X	X	X	X	X
Renovation	X	X			X
New Construction	X	X			

In buildings designed to employ natural illumination, and with the proper wiring of the supply power, daylighting can be used.

Dimmable solid-state ballasts are best used in renovation and new construction because a ballast must be installed in each fixture. In addition to the efficacious operation of the fluorescent lamp system, ballasts allow four major lighting control strategies. The advantages, as compared to the other controllers, are increased dimming range and the control of light levels via low-voltage signals. For buildings designed to exploit natural light, five strategies can be accomplished with solid-state dimming ballasts.

The strategies vary when using clocks, sensors and photocells. Clocks must indicate real time for employing the predictable scheduling strategy. Photocells are needed to measure ambient illumination levels for accomplishing lumen depreciation or the daylighting strategy. For lumen depreciation, only a few photocells are required. Considerably more photocells must be used for daylighting because the dynamic nature of natural illumination requires measuring illumination levels over smaller areas. Personnel sensors are most applicable to retrofit situations in spaces that are occupied intermittently during the day. They are the only devices that can automatically respond to unpredictable occupancy of a space.

Communications Controls. The communications central information processors and power-line carriers are auxiliary controls used in conjunction with lighting controllers and sensors.

They connect the instructions with the commands. The central processors are most effective with the scheduling, load shedding, and lumen depreciation strategies.

Transmitting information over power lines eliminates the need to string wire. In retrofit applications, rewiring could make the control technique too costly, so power-line communication is most attractive. In renovation and new construction, where rewiring will be done, power-line carrier methods are less attractive because of installation costs.

APPLICATIONS AND COST FACTORS

Decision-makers are faced with three types of applications choices (retrofit, renovation, or new construction) for a lighting control system. Table 5-3 shows the five major factors to be considered for each type of application.

A retrofit replaces or adds to an existing lighting system that is already operating adequately. The primary objective of a retrofit is to reduce operating costs. It is unlikely that a retrofit application would be economically sound if the supply circuit layout or the building design would have to be altered. The ideal retrofit requires no hard-wiring to the supply line. The above arguments are the reasons all five major cost factors must be considered for a retrofit.

When a building is renovated, the entire lighting system is replaced as well as all the supply circuit wiring. Thus, for renovation applications, installation costs and supply circuit layout are no longer major factors. That is, they would have been replaced at some cost in any case. The three major cost factors for a renovation are the initial and operating costs and the maintenance of the building design. This means the lighting control system can employ more control strategies since there are fewer application constraints.

New construction is an application where the lighting control system can influence the building design. For example, many new buildings are designed to optimize the use of natural illumination. The building structure, position, and fenestration system are designed to accommodate the lighting control technique.

Thus, the building design is not a major cost factor in new construction applications. This permits the use of daylighting. As shown in Table 5-2, a greater number of lighting control strategies are feasible in new construction.

APPLICATION AND EQUIPMENT

Table 5-4 lists the lighting control equipment that best suits each type of application. For each application, there are several options, which are optimum for different situations. The table lists control equipment in groups of priorities for each application. In addition the number of strategies that are optimum for each type of control are indicated in parenthesis. Priority I includes the equipment that meets all the conditions for the particular application. For example, delamping in a retrofit application has no initial cost, saves 50 percent in operating cost, has very small installation costs (removing the lamp and disconnecting the ballast from the supply), requires no change

Table 5-4. Equipment Selections for Various Applications

Priority	Retrofit	Renovation	New Construction
I	(1-2) Delamp	(2) Switches/Relays	(2) Switches/Relays
	(1-2) Impedence-Modifier (lamps)	(3) Voltage/Phase Control	(4) Voltage/Phase Control
	(1) Switches/Relays	(4) Solid-State Dimming Ballasts	(5) Solid-State Dimming Ballasts
	(1-2) Voltage/Phase Control	(3) Computer/Micro-processor	(4) Computer/Micro-processor
	(1) Computer/Micro-processor	(1) Clocks	(1) Clocks
	(1) Power-Line Carrier	(1) Photocells	(2) Photocells
	(1) Clocks		
II	(1-2) Impedance-Modifiers (hard-wired)	(5) Power-Line Carrier	(5) Power-Line Carrier
	(1) Personnel Sensors	(1) Personnel Sensors	(1) Personnel Sensors
	(1) Photocells		
	(3) Voltage/Phase Control		

( ) Numbers in parentheses indicate the number of lighting control strategies that can be implemented for an application.

in supply circuit layout, and does not require a change in building design.

The group of retrofit equipment classified as Priority II involves significant installation costs and requires stringing wire throughout the floor. These satisfy four of the five major cost factors.

The power-line carrier is in the Priority I group for a retrofit and Priority II in renovation and new construction applications. This is because layiing control lines is not required when retrofitting. In renovation or new construction, the expense of laying the control lines is much less—that is, installation costs are not a major factor.

Personnel sensors are also rated Priority II for renovation and new construction. Although installation cost is not a major factor, the uncertainty of the activity in the space affects operating costs.

The difference in the number of strategies that can be accomplished in a renovated space and in new construction is shown in Tables 5-2 and 5-4. The additional strategy that can be employed in new construction is daylighting. This does not imply that daylighting is impossible in renovation applications. Daylighting depends on the building design, which is a major cost factor. However, some existing buildings are suitably designed to use daylighting effectively.

SUMMARY

A general approach for selecting the lighting control equipment that best suits an application is based on the interdependence of equipment, control strategies, major cost factors, and applications. The equipment listed for each group is not exclusive; there could be circumstances in which equipment or a strategy not listed would be suitable. More importantly, this chapter shows that a decision-maker has several options for lighting control systems for any application.

Chapter 6
Selecting Motion Sensor Technology For a Large-Scale Lighting Project

Jerry D. Andis, C.E.M.
TRW INC

The successful implementation of a large-scale lighting control project at the Redondo Beach complex of TRW Space and Defense revolved around ensuring employee satifaction and technical performance. The combination which resulted in overall project success is described in this chapter, including all phases of the project approach from initial test, Phase I, evaluation of motion sensor technologies, and a detailed description of the various steps of final implementation (Phase II).

The project involved evaluating and recommending a cost effective approach for achieving lighting control in 27 buildings with a gross square footage exceeding 3 million. Phase I of the project involved the installation of 550 sensors in a pilot program. After Phase I acceptance, a detailed evaluation was performed and Phase II was implemented with the installation of approximately 4600 sensors.

PROBLEM STATEMENT

The TRW Space Park complex in Redondo Beach, CA, consists of approximately 27 buildings arranged in a campus-type setting. The majority of the buildings were constructed in the mid to late sixties. Employee population at Space Park is approximately 7500. Building use is varied, consisting of manufacturing, engineering and development offices, and computer facilities.

77

The building lighting was previously controlled manually through central panels. Each building (typically two stories) has 8 lighting panels located in the corridors. The majority of the buildings contained no light switches in the individual offices. Only the offices which had been modified over the years and the conference rooms contained light switches. The lighting prior to the project was controlled by building security personnel. Security shut off lighting panels after cleaning personnel were completed with their work, typically 2 a.m.

Energy monitoring of lighting panels indicated burn time on the average of 22 hours/day. This was the driver for improved lighting control. A new method was needed that would be flexible, reliable, and most importantly, responsive to individual behavior.

LIGHTING ALTERNATIVES

Three basic alternatives were considered. The first alternative was central lighting control via an existing energy management system. This alternative proved to be expensive and could not provide the needed flexibility. At the time of the evaluation, Space Park did not have touch-tone phones which would have provided a flexible on/off function. The second alternative was to install standard light switches. This alternative was not attractive due to cost and dependence on employees to utilize switches. The third alternative was the use of occupancy sensors, either Passive Infrared or Ultrasonic. The ultrasonic occupancy sensor was selected after a detailed evaluation.

Occupancy sensors would provide responsive and selective lighting control. This was an important feature considering the varied working environment at TRW where the flexible hours available for employees and different functional areas, such as manufacturing, drive a variety of work hours.

PHASED APPROACH

A three-phase project approach was selected: Testing, Phase I, and Phase II. The testing phase consisted of timer tests with seven units in a standard office bay. A standard office bay consisted of a secretarial bay surrounded by six offices (refer to Figure 6-1). Two timers were installed. Timer A recorded the hours that the room was occupied, and timer B recorded the on-time of the lighting load. The difference between the readings is the wasted lighting. The results from the timer tests ranged from 33-69 percent with an average of 55 percent.

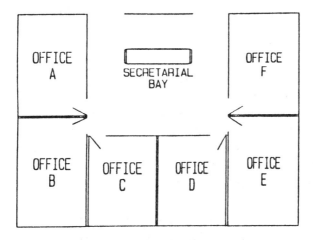

Figure 6-1. Standard Office Bay

The savings range variance resulted from occupancy differences between the offices.

The product selected for testing was the Light-O-Matic, an ultrasonic motion sensor manufactured by Novitas. The particular product selected was not critical during the testing phase. The testing phase documented savings and evaluated employee satisfaction.

In Phase I of the project, 550 units were installed in a pilot program in offices and conference rooms of the first floors of four different buildings. The product selected was the same as the testing phase due to performance and reliability of the product. The purpose of Phase I was to perform a larger scale test for further evaluation of product performance, occupant satisfaction and documentation of energy savings. An employee satisfaction questionnaire was distributed to evaluate and provide information regarding any dissatisfaction and confirm correct operation of sensors.

The survey results showed that 75 percent of the employees rated motion sensors as above average or excellent and 25 percent responded that sensors were inconvenient since units switched off while they were in their offices.

Following the questionnaire an adjustment cycle on sensor sensitivity was performed. The offices with computers needed a higher sensitivity. This was due to minor motion in the office as employees worked on personal computers.

The adjustment cycle eliminated a majority of the problems. To further verify energy savings against the initial test data, selected lighting panels were monitored for a period of two weeks. The energy recorder measured kW demand on a 15-minute interval. The analysis of the data showed a 47 percent savings on the monitored lighting panels (Figure 6-2). For project justification a 50 percent savings was used by averaging initial test data with Phase I savings.

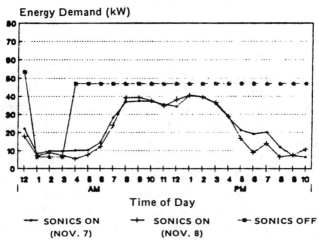

Figure 6-2. Lighting Panel Monitoring

A 50 percent savings yielded up to 11 hours per day (approximately 3000 hours per year) of wasted lighting (Figure 6-3). The majority of the savings occurred between the hours of 6 p.m. to 2 a.m., due to building janitorial service.

Phase I provided data for implementing Phase II. To complete the total complex, including the addition of planned new buildings, quantities were estimated at 6800 sensor units. This required a request for about $1.4 million in capital and $70,000 in expenses. The resultant annual energy savings estimated based on 50 percent savings was approximately $1.35 million. Simple payback (including energy tax credits) was .9 years. The internal rate of return for the project was calculated at 61.1 percent. The TRW guideline for evaluating energy management projects is a simple payback of 2.5 years or less and an internal rate of return of 15 percent or greater.

WASTED LIGHTING IN OFFICES

Period	Lighting Hours	Wasted Hours
8 AM – 12 PM	8.0	3.0
12 PM – 2 PM	2.0	0.5
2 PM – 6 PM	4.0	0.5
6 PM – 2 AM	8.0	7.0
Daily Hours	22.0	11.0
% of Daily Hrs.	100%	50%

Figure 6-3. Wasted Lighting Hours

PROJECT IMPLEMENTATION

During the management approval cycle of project funding, a detailed analysis was performed on various ultrasonic product parameters. This comparison and evaluation was used as the basis for Phase II ultrasonic product selection. Comparison of ultrasonic products was performed across 13 different categories. The products were divided into high, medium, and low categories and weighted accordingly to arrive at a numerical score.

Table 6-1 describes the 13 parameters evaluated for product selection. Parameters in the high category were the main drivers in product selection. The most important parameter was product performance and reliability since workers' productivity carries the highest priority at TRW. Therefore the product selected must not inconvenience or negatively impact the working environment. The second most important parameter was the selection of a firm which could provide support services needed for a project of this magnitude. These services included working with the contractor to determine correct sensor locations in the work space and most important, followup adjustment services. The followup adjustment service was critical to ensure total occupant satisfaction with the new occupancy sensors.

The initial product selected was manufactured by Novitas, Inc., Santa Monica, CA, and distributed by Intecorp, Santa Ana, CA. It was rated as the best for the large-scale installation at TRW. The product was the Novitas ultrasonic motion sensor, Model 01-071, along with its associated relay and transformer.

Table 6-1. Product Comparison Parameters

PRODUCT SELECTION	MEDIUM VALUE
	Quality of Components
Ultrasonic Manufactor	Application Coverage
Comparison Parameters	Product Performance
	By-Pass Function
HIGH VALUE	
	LOW VALUE
Protect Performance & Reliability	
Support Services	Installation Procedures
Competitive Prices	Circuit Board Protection
Warranty Terms	Defective Unit Replacement
	Operating Frequency
	Operation Costs

PRODUCT PROCUREMENT

In determining a cost effective approach, a direct buy vs. an electrical contractor purchase of material (occupancy sensors) was evaluated. It was determined that a TRW direct purchase from the distributor was most cost effective. Advantages of the direct buy included a large quantity discount and control over product delivery schedules. The total quantity (6800 sensors) was delivered in three shipments over a 4-month period.

APPLICATION CONSIDERATIONS

The typical office is 120 sq ft. The sensor selected covers 900 sq ft (½ step motion) and 670 sq ft (minor motion). To avoid ultrasonic interaction, such as ghost turn-on, among the office bays, secretarial areas utilized a sensor on a different frequency than surrounding offices. The standard office bay with hard walls, tile floors, and office doors provided too much reflective surface for all sensors to operate on the same frequency. Installing frequency B sensors in the main secretarial bays eliminated the problem.

Another application issue was determining the correct sensitivity adjustment for the various offices. The standard office, without a personal computer, was best suited to a sensitivity adjustment of approximately 1-2 in a range 1-10. Offices containing personal computers required that the sensitivity be adjusted up slightly (4-5). The timing feature was initially set for 7-8 minutes in all offices. However, it was determined that due to the nature of the work, typically engineering offices, that a time period of 12 minutes in a range .5-12 minutes would be more suitable. The time delay of 12 minutes was selected to minimize the potential of any false turn-offs. As stated before, the highest priority during the project implementation was given to not impacting the workers' environment or productivity.

CONSTRUCTION CONSIDERATIONS

The bid forms requested that each contractor provide unit pricing based on the wiring type as determined in the design. A firm fixed price was based on selection of the most common wiring type as determined by TRW. All contractors bid on a firm price to install 6800 units. This allowed an evaluation of the competitive bid and provided the flexibility needed, as the exact sensor or specific wiring type counts were not known at bid time. The final count of installed sensors during Phase II was approximately 4600 units in a period of less than five months. The electrical contractor installed an average of 50 sensors per day.

The installation was done between the hours of 5 p.m. to 2 a.m. to minimize the impact on employees. The contractor was instructed to initially set the sensors based on directions from the distributor. This eliminated a considerable amount of initial complaint calls due to improper sensor settings. A quality control program was also stressed to the contractor to ensure that any cross wiring was kept to a minimum. The cycling of the units several times prior to leaving the office improved the chances of discovering potential failures during the initial installation. Due to the rate at which the project moved, it was essential that a good quality control program was maintained to ensure smooth project implementation.

The total project cost for the installation of 4600 units was approximately $850,000. This equates to an installed per unit cost of $186 and a verified simple payback of 1.1 years. The keys to the successful implementation of this project are given below.

Keys to Successful Implementation

The following are key points to ensure successful implementation of a large-scale occupancy sensor project:

1. Perform testing via timer tests and document savings for project justification.

2. Evaluate product and services provided by distributor.
   - Establish parameters to evaluate
   - Evaluate vendor services
     - Sensor location layout
     - Followup adjustment

3. Minimize impact on employees' working environment.
   - Notification flyer prior to construction.
   - Immediate response to complaints
   - Followup questionnaire to ensure satisfaction with sensor operation
   - Perform final adjustments based on occupant feedback

4. Maintain and monitor project after installation.

Chapter 7
California Title 24 Lighting Controls Encourage Automated Control Technologies

David Peterson

The California Energy Commission has targeted lighting as a key load in setting up new construction and renovation criteria. The approach is twofold: limit the connected lighting load, and insist on minimum control capability while encouraging the use of more advanced technologies.

Limiting the connected load provides a firm foundation for insuring savings. By reducing the watts/sq ft, you are essentially guaranteed that the owner is going to reduce both his electrical demand and consumption.

To wring the maximum savings out of this strategy, Title 24 power budgets are set to make the lighting designer stretch to meet the codes and still provide a good seeing environment. High efficiency sources and nonuniform layouts may be required, and less efficient incandescent sources and "aesthetic" lighting may have to be deleted.

But you can only go so far reducing the watts/sq ft and who wants to work in a bland space? Automated lighting controls offer an attractive alternative. Numerous studies have shown major reductions in both consumption and demand through their use. Title 24 encourages the use of these automated control technologies without mandating them (Figure 7-1). Minimum control capabilities, however, are mandated. Basically, these minimum controls allow the conscious user to manage his lighting power and consumption manually. They reflect the need for an individual to be able to get the lights ON in his area and to reduce the lighting if adequate daylighting is available. However, if the designer opts for automated forms of control which also meet the need to reduce the lighting for daylighting, he is given relief on his lighting power budget.

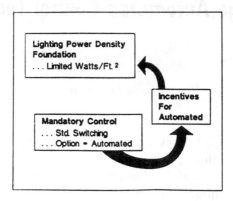

Figure 7-1. California Title 24 Lighting Approach

LIGHTING CONTROL FUNCTIONS

The whole purpose of lighting control is to get the right amount of light, where needed and when needed.

Occupancy Based Control. The single most powerful strategy is to match the lighting to occupancy. The two general approaches used are scheduling with local override and occupancy sensors. Scheduling may be simple ON/OFF or the lighting levels may be programmed to match the anticipated area functions. For instance, full lighting during normal workhours, reduced lighting for cleaning, startup and shutdown. For occupancy-based scheduling to work, however, it has to allow local occupant override. Occupancy sensors eliminate the need for scheduling. Instead, the lights are turned ON automatically whenever someone enters the space.

Daylighting. Automatic daylighting controls attempt to maintain uniform lighting levels on the task by adjusting the output of the lighting fixtures. This adjustment can take the form of switching of selected fixtures or lamps or dimming. One word of caution—it isn't easy, and it isn't cheap. A good daylighting job requires an integrated design approach encompassing the building architecture and controls. Title 24 encourages daylighting in principle, but it always requires occupancy-based controls in addition.

Tuning and Lumen Maintenance. Tuning and lumen maintenance are dimming-based strategies. Tuning allows the lighting output of individual fixtures or zones to be adjusted to match the particular needs of the occupants in that space. For instance, fixtures in traffic areas can be adjusted down and then returned to full output if the layout should change. Lumen maintenance simply attempts to save lighting energy by automatically compensating for the overlighting designed into a layout to compensate for lamp and fixture deterioration.

Management Data and Building Automation. Lighting automation may also provide management data. For instance, after-hours usage may be billed directly to the responsible tenants rather than uniformly distributed to all; central graphic displays can alert security to unauthorized use of spaces, and printouts of cleaning lighting overrides can be used to track cleaning crews.

TITLE 24 MANDATORY CONTROLS

Title 24 requires that all new office construction meet minimum requirements as spelled out in Sec 2-5319. Basically this section provides for multiple level local switching and daylight circuiting "where effective." Programmable controls and occupancy sensors provide acceptable alternatives to hardwired switches, but the daylight switching conditions still must be met.

Enclosed Space. Sec 2-5319 requires that any space with floor to ceiling walls greater than 100 sq ft with a connected lighting load greater than 1.0 watts/sq be provided with multiple level switching capable of providing uniform reduced illumination (see Figure 7-2). In addition, the controlled area must be readily visible from the switching location.

ENCLOSED SPACE   >100 FT²
                  >1.0 WATT FT²
                          MULTILEVEL SWITCH

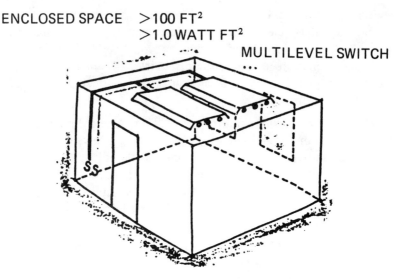

Figure 7-2. Enclosed Spaces

Open Areas. There's no limit on the size zone which can be switched by a single device. Theoretically, you could have two contactors control 10,000 sq ft of space in an open area but this is a misapplication. You should plan on a maximum switched zone size of 1000 sq ft (see Figure 7-3).

NON-DAYLIGHT VS. DAYLIGHT

The code mandates that local controls take into account daylighting. Specifically "lighting circuiting shall be arranged so that units, in all portions of the areas where natural light is available at the same time, are switched independently of the remainder of the area." In practice, this means that you would wire perimeter zones or offices to switch the lamps in order to maintain uniform lighting when daylighting is available. For example, you could switch the fixtures next to the windows entirely or partially as indicated for the three-lamp configuration (see Figure 7-4).

OPEN AREAS . . . _?_ FT² MAX.

. . . SWITCHED/"XL"

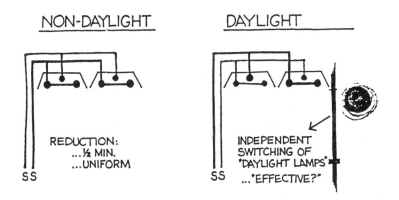

PLAN ON 1000 FT² MAX.

Figure 7-3. Open Areas

NON-DAYLIGHT

DAYLIGHT

REDUCTION:
...½ MIN.
...UNIFORM

SS

INDEPENDENT
SWITCHING OF
"DAYLIGHT LAMPS"
..."EFFECTIVE?"

SS

Figure 7-4. Non-Daylight vs. Daylight

The code doesn't specify the percentage of the lighting that must be switched for the daylight configuration. The implication is 50 percent. Specifications for large open zones are fuzzy. Strict adherence to the code would imply a minimum of three switches for the zone, two to provide the multi-level switching and the third to independently switch lamps in the daylit area. The remainder of the lamps in the daylit zone could be switched with one of the other two switches.[1]

Programmable Relay System

The code recognizes that a programmable relay or occupancy sensor are equivalent to dual-level manual switching, but once again we must take into account daylighting where "effective."

Non-Daylight Application. For instance, in a non-daylit zone, the relay can take the place of the two switches directly. The intelligence driving the relay, however, must meet certain minimum requirements. First, it has to provide at least two different schedules (one for weekdays and the second for weekends). Furthermore, local override of the relay must be accommodated. This could be in the form of a local switch or telephone. This override can not permanently disable the automatic schedule; that is, the override must be cleared and the automatic schedule restored the next time around.

Daylight Application. If there is effective daylighting in the area, manual switching of the daylight circuit must also be provided. For a programmable system, this would mean an extra relay to handle the daylight circuit. The code *doesn't* require that the daylight relay be controlled automatically, but simply that the circuiting and switching mechanism be provided. Automating the daylight relay provides additional credits, but it is not required.

Occupancy Sensors

Non-Daylight Application. When you use an occupancy sensor in a non-daylit zone, no other local switching is required.

---

[1] The space that received daylight is referred to as a "daylit" area or zone.

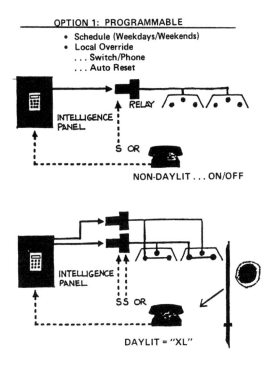

Figure 7-5. Programmable Relay System

Daylight Application. But once again, if the area has "effective" daylighting, you must supply switching of the daylight associated lamps. The "gotcha" here is that elimination of the switches helps pay for the incremental hardware cost of the occupancy sensor. Daylight compensation forces you to stick a local switch back in.

CONTROL INCENTIVES

Not only are programmable relays and occupancy sensors approved alternates to manual switching, incentives are provided for their use. These incentives are in the form of a percent increase into the code limit on watts/sq ft for each space controlled. The amount of the credit varies with the usage of space controlled and the type of controls used. For example, using

OPTION 2: OCCUPANCY SENSOR

NON·DAYLIT

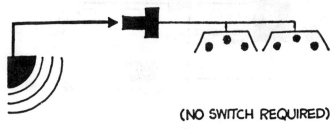

(NO SWITCH REQUIRED)

DAYLIT = "XL"

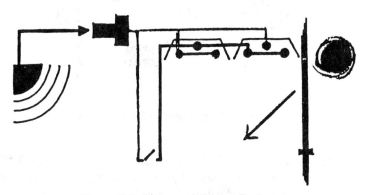

Figure 7-6. Occupancy Sensor System

an occupancy sensor in an individual office increases the allowable power density 30 percent, or 1.3 times the normal 1.5 watts/sq ft. In an open area, the increase would be 15 percent, reflecting the reduced effectiveness of the occupancy-sensing function in open areas.

In the daylit areas, the occupancy sensor and the programmable system had to provide for manual switching of the "daylight" fixtures. The manual control capability is mandated. If you automate the function, the power credit is increased. For instance, in the office (see Figure 7-7) the two relays in the

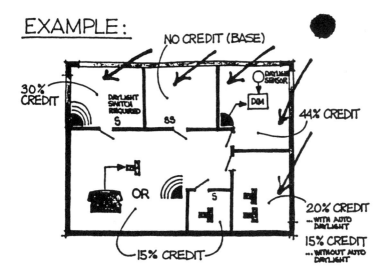

EXAMPLE:

NO CREDIT (BASE)

30% CREDIT

DAYLIGHT SWITCH REQUIRED

S

8S

DAYLIGHT SENSOR

DIM

44% CREDIT

OR

S

15% CREDIT

20% CREDIT ...WITH AUTO DAYLIGHT

15% CREDIT ...WITHOUT AUTO DAYLIGHT

Figure 7-7. Power Credits

lower right corner provide the necessary switching of the daylight and remaining fixtures by providing an automatic daylight sensor with associated software. With this arrangement, the credit is increased from 15 percent to 20 percent.

Be wary of scale factors. For instance, in a typical office 60 percent of the floor is normally open space; 40 percent, individual offices. An individual office would normally be about 200 sq ft, with the open zone, 1000 sq ft. This means that one programmable relay point in an open zone will deliver five times the credit as the same control point in an individual office. This "scale factor" has tremendous impact on the economics.

COST/PERFORMANCE TRADEOFFS

Let's take a closer look at the cost/performance tradeoffs of the different approaches: hardwired line-voltage switching, programmable relays, and occupancy sensors.

Individual Office Cost. In examining the individual office space, remember that there are two major conditions to be met—daylight and non-daylight. The daylit spaces require the multiple level (XL) control for both occupancy sensors and programmable relays.

Simple ON/OFF control with a switch is not allowed, but it is included for comparison purposes. Hardware costs are low, about $20. Bringing line voltage switching down the wall is an expensive proposition. Total costs for a single switch, $85; for two (XL), $105 (see Figure 7-8).

The programmable ON/OFF doesn't look too bad at $130. The daylit space, however, forces the use of two control points and the $205 per office reflects the impact.

The same problem shows up with the occupancy sensor. Again, the straight ON/OFF control looks pretty good at $150, but throwing in a hardwired switch for daylight control pushes you well past the $200 per office mark.

Zone Cost. Control of open areas or zones is no different than individual offices for switches or programmables. However, occupancy sensors for open spaces need more sophisticated sensing capability. This is reflected in a much higher cost per point. Here the economics clearly favor programmable or standard switching (Figure 7-9).

Performance. But cost is only one side of the equation. The energy side is reflected in the control credits. Although the relative magnitude of savings potential are debatable, both occupancy sensors and programmable relays outperform standard switching.

However, other performance factors also need to be taken into account. For instance, the typical occupant couldn't care less about energy savings. He simply wants to control his own lighting. From his perspective, switches are just fine. Any form of automation is an intrusion, but providing overrides for the programmable system and eliminating nuisance switching with the occupancy sensors make both acceptable. For the building operations personnel, management data can be more important than energy savings. That's quickly evident when you talk to a building owner who charges overtime usage on a pro-rated rental basis.

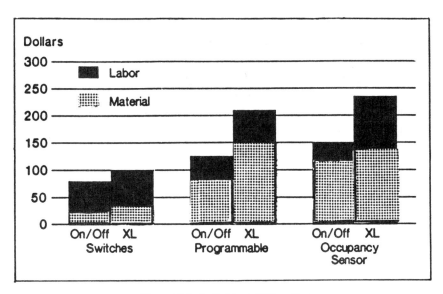

Figure 7-8. Individual Office Cost

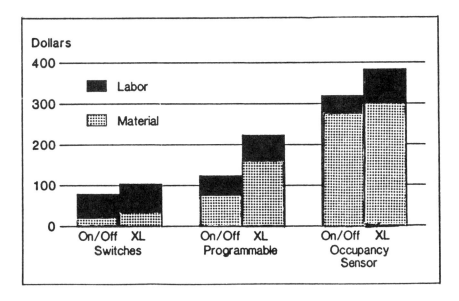

Figure 7-9. Zone Cost

CONCLUSION

Taking all factors into account, individual control is a close call. Daylight switching requirements in most individual offices throw a curve at the automation options. Open area or zone control, on the other hand, is a clear "gimme." Programmable relay based systems provide the savings and credits at very attractive costs.

SECTION III
REFLECTORS

SECTION III

Chapter 8
Fluorescent Reflectors: The Main Considerations

Paul von Paumgartten

The benefits of replacing lamps in fluorescent light fixtures with reflector panels have been well documented. Articles have touched on everything from energy savings and reduced lamp and ballast maintenance expenses to lower air conditioning costs, extended lens life and reduced glare. These benefits have helped make reflector retrofitting the fastest growing segment of energy management.

At a glance, it appears simple enough. Investment in the retrofit and reap the return. But, in choosing to retrofit a building's fixtures, building owners and managers must closely examine a variety of considerations that will affect the budgetary savings, as well as the overall satisfaction of the building occupants. This chapter will spell out those technical factors that building managers and owners should evaluate when considering a fluorescent fixture retrofit plan.

Among the key considerations are:

- *Light Losses.* Replacing lamps with reflectors decreases light levels, but they may be maintained within current lighting standards and not adversely affect worker productivity.

- *Fixture Appearance.* The apperance of the fixture is critical, because people respond to lighting emotionally, as well as physically.

- *Designs.* The styles vary from full reflectors to partial, with varying effects on the lighting system.

- *Material.* Each has its own reflective qualities, affecting reflector output, appearance, and durability.

- *Costs.* The various reflective materials and designs present a cost-benefit dilemma.

- *Heating, Ventilating and Air Conditioning System Changes.* Fewer lights mean less heat, reduced cooling load and increased load.

LIGHT LOSSES

Logically, there is light lost when a lamp is removed and reflectors are installed. If two lamps from a two-by-four-foot, four-lamp troffer are replaced with optical reflectors, light losses may be 10 to 40 percent.

But, things are not always as they appear at first glance. Light losses vary from project to project depending on the fixture type and condition. Dirty lenses or old lamps cause fixtures to operate well below their designed efficiencies; they are dim to begin with. Buildings with poor lighting maintenance may be retrofit with reflectors with very little, if any, difference in light level. The illuminance that would be lost in an average reflector installation can be made up by cleaning the fixture and replacing the old lamps with new ones during the installation (see Figure 8-1). Installers can do this during a retrofit.

Figure 8-1. Reflector Charts

Fixture Efficiency	
Fixture Condition	**Efficiency**
Seasoned and cleaned white troffer with new lamps and reflectors. (Retrofit fixture)	65-85%
Seasoned and cleaned white troffer with new lamps. (New fixture)	50-65%
Seasoned and uncleaned white troffer with seasoned lamps. (Old fixture)	35-45%

That's what makes the reflector concept work. Generally, a 10 to 25 percent light loss can be expected with a reflector retrofit.

In addition, the light loss may not be significant to the needs of the building occupants. Before the move to conserve energy, many building designers over specified for lighting. That may leave room to reduce lighting without cutting productivity.

As it is, new lighting standards suggest lower lighting levels are acceptable. Previously, most buildings were designed for light levels between 70 and 100 footcandles (FC). But, according to the Illuminating Engineering Society of North America lighting recommendations, offices need only 30 to 50 FC for Video Display Terminal usage; 50 to 70 FC for reading, writing, and typing; and 70 to 100 FC for accounting and drafting. In general lighting areas, offices need 10 to 20 FC for circulation areas, corridors and lobbies; 25 to 35 FC for conference rooms and non-task areas; and 30 to 40 FC for filing areas (see Figure 8-2). Task lighting can meet the needs of workers in many areas, while lower level ambient lighting is sufficient elsewhere.

Figure 8-2. Recommended Light Levels

Task Lighting Requirements	Footcandles
VDT Usage	30-50
Reading, writing, typing	50-70
Accounting, drafting	70-100
General Lighting Requirements	
Circulation areas, corridors, lobbies	10-20
Conference Rooms, non-task areas, work stations	25-35
Filing areas	30-40

Previously, building designers used 2.5 to 3 watts per square foot for lighting. Today, the standards have been reduced to 1.5 watts per square foot—without necessarily reducing occupant satisfaction or employee performance.

In addition to loss of light, reflectors can change the light distribution from a fixture. Reflectors direct light downward, creating more light directly under the fixture and less light between fixtures.

This light redirection is similar to that of the low-brightness louvers, which eliminate the high-angle light. This brightness control can reduce glare in offices where computer video display terminals are in use (see Figure 8-3). Spacing requirements may change because of less light between fixtures.

Figure 8-3.

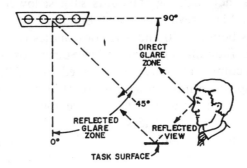

If the two-by-four-foot, four-lamp fixtures are providing more light than necessary, redesigning the fixture for reflectors could be readily acceptable to office workers. It is also less expensive than installing a new lighting fixture completely. But, several other key conditions remain in selecting the appropriate type of lighting designs using reflectors.

FIXTURE APPEARANCE

Aesthetics, though difficult to quantify, are very important to a building environment. Many are obsessed with how much light is produced from a fluorescent reflector, but they may be over looking an important factor: the overall acceptance of a retrofit by the building occupants. Appearance can make a big difference in working conditions and worker satisfaction.

That is why delamping alone is not recommended for energy savings in an office setting. Delamping leaves a dark spot in the fixture where the lamps have been removed. This is aesthetically unacceptable to building occupants.

It has been shown that a great looking retrofit will more often prove to be successful than a bright one. Ambient lighting from overhead fixtures, task lighting for specific jobs or even accent lighting for visual affect must be done in an appropriate, tasteful manner.

DESIGNS

The reflector design can have an impact on aesthetics. As a minimum, a good looking retrofit fluorescent fixture should do one of two things: Create an image that the fixture contains a full complement of lamps or diffuse the light so no lamps can be seen in the fixture whatsoever. Different reflector shapes and materials will yield these results.

Some designs involve moving the lamp holders and repositioning the remaining lamp holders and repositioning the remaining lamps. Some virtually line the inside of the fixture with reflective materials. Others involve simply replacing existing lamps with reflectors. The reflectors may replace the in-board or the out-board lamps in a four-lamp fixture. Some have bends that help create the image of a lamp where one has been removed. And some are partial reflectors, which leave the ballase uncovered for easy access (see Figure 8-4).

Figure 8-4.

Standard troffer with 4 lamps.

Simple bend reflector with 2 lamps outboard.

Complex bend reflector with 2 lamps inboard.

Complex bend reflector with 2 lamps relocated.

Partial reflectors with 2 lamps outboard. Ballast cover is not covered.

MATERIALS

There are several different materials that can be used. Their reflective qualities differ, with silver film on aluminum rating the highest with a 90 to 95 percent reflexivity (see Figure 8-5). From a building occupant's perspective, however, the difference between materials may be imperishable. Depending on the reflector design, the material used may have little effect on light output.

Figure 8-5. Material Reflectances

Material Types	Total Reflectivity
Silver film (on aluminum)	90-95%
Aluminum film (on aluminum)	80-85%
White paint (on aluminum)	75-90%
Anodized aluminum sheet	70-85%

Perhaps the best way to get a true impression of how a reflector design will look is to perform a mock-up. Have it installed in the ceiling to see the actual fixture where it will actually be used and measure the performance. Do not compare the measurements to the light levels that existed previously; instead, compare them to current lighting standards. The key question is: Will the reflector yield an acceptable light level?

Some types of reflectors will change in appearance over time. While aluminum reflectors have been proven durable in use for over 30 years, the thin anodized coating on aluminum reflectors—generally about two microns—may crack when the reflector is bent, causing possible oxidation of the aluminum.

While aluminum films have a long history, silver films have little track record. But research groups have run accelerated aging tests on silver films that show films with back reflector materials age faster, losing more than 20 percent of their reflectivity over 500 hours. Front reflectors maintained their reflectivity through the course of the study.

In time, temperature swings in the fixtures can also cause silver film to crack or buckle. Some film manufacturers, however, have added "ultraviolet inhibitors" to counter the effects of temperature swings.

Over time, a greater attention will have to be paid to lighting maintenance following a retrofit. Considerations should be made to group relamping on a regular, timely basis.

COSTS

While the effects of reflector designs and materials may have considerable impact on appearance, the retrofit's costs will have the greatest bottom-line impact. A high-cost retrofit could double the payback period.

The costs of the materials can vary greatly. Silver film reflectors are about $10 more expensive per fixture than aluminum. The average cost for a silver film reflector for a two-by-four-foot fluorescent fixture is $50 to $75 installed, compared to $40 to $65 for aluminum. Partial reflectors with aluminum film cost between $35 and $45 (see Figure 8-6).

Figure 8-6. Cost Comparisons

Reflector Type	Typical Installed Cost	Payback*
Silver Film (Full)	$50-$75	2.3-3.4 yrs.
Anodized Aluminum (Full)	$40-$65	1.8-3.0 yrs.
Aluminum Film (Partial)	$35-$45	1.6-2.0 yrs.

*Based on 90 watts saved times 3,500 hours/per times seven cents/kWh or $22.05/year.

Installation accounts for 25 to 40 percent of the cost. But lowest installation price may not be the best option, since proper installation is critical to the success of the retrofit. The installation contract should include new lamps and a thorough fixture cleaning to minimize the light loss in the retrofit.

The installation should also be done when it will least disrupt the building occupants. If performed after hours, the lamps can be removed and reflectors installed without the building occupants realizing there has been a change.

HVAC IMPLICATIONS

Improper installation or design can also mean additional hidden costs. Many light fixtures serve as return air ducts, an integral part of the heating, ventilating and air conditioning system. Full reflectors may cover the vents, impeding proper operation. Partial reflectors can render the necessary levels of illuminance without interfering with the air duct systems.

Also, building managers may not want to cover the ballast compartments. That increases the compartment temperature and shortens the life of the ballast. It also makes it more difficult to reach the ballast for maintenance procedures.

There are several considerations in HVAC systems that can drastically affect a building operation expenses as well as occupant comfort.

Reduced lighting causes a corresponding reduction in the cooling load for the air conditioning equipment, especially in the interior zones, where outdoor conditions have little influence. Demand for winter space heating may increase incrementally with reduced building lighting. This decreases the savings from the light reduction program by the amount of energy that must be added to offset the loss of heat. However, a building's heating system generally provides spare heat much more efficiently than the heat given off by excessive lighting.

For example, in a terminal reheat system, a change in lighting could require as much additional energy to reheat the duct air as is saved by reducing the lighting. The reheat requirement, however, can be minimized by raising the cool supply air temperature so comfort conditions in the room with the maximum cooling load are satisfied without reheating the air going to other rooms.

In the variable air volume system, a reduced cooling load would reduce the amount of cool air that is distributed through the building. This reduction may present an opportunity to replace the supply fan motors with smaller motors, saving additional energy. An HVAC expert is necessary to evaluate the retrofit savings potential.

A reflector retrofit also presents an opportunity to change the entire lighting system. Different lamps for better color rendition or greater efficiency could be installed. Parabolic louvres replacing the old lenses would reduce glare on office Video Display Terminals. And high-efficiency ballasts would afford even greater energy savings.

SUMMARY

With lights burning between 30 and 50 percent of a building's electrical load, many building owners and managers have chosen to install reflectors in their fluorescent fixtures as a quick-fix way to cut energy expenses. But the decision to install reflectors should involve more than simply changing a fixture. Aesthetics, designs, materials, costs and the retrofit implication to the HVAC system all need to be considered carefully. It also gives building management an opportunity to redesign the entire lighting system.

Careful selection of a reflector system and installer can help building management meet its desires for energy savings while increasing the satisfaction and productivity of building occupants.

Chapter 9
Specular Retrofit Reflectors For Fluorescent Troffers

J.L. Lindsey

INTRODUCTION

Specular optical reflectors for fluorescent fixtures can, according to some manufacturers' claims, permit the removal of 2 lamps from a 4-lamp troffer and actually result in an increase in lighting levels. Other manufacturers state that, in general, a reduction in maintained illuminance will occur but that the reduction is acceptable in many cases. These conflicting claims have generated substantial interest among consumers and members of the lighting community. Who is right and who is wrong? Many questions are asked, and the issue is clouded by lack of reliable information or accurate, detailed analyses.

This chapter will explore the performance characteristics and economics of several reflectors designed for use in prismatic lensed troffers, and their ongoing effects upon the performance of the lighting systems into which they are incorporated. Reflectors for troffers employing louvers, and for strip fixtures commonly used in industrial and warehousing applications, are not included in this analysis. These reflector applications will be the subject of a future paper which will be prepared when sufficient data become available.

The photometric data used for the analysis were prepared by recognized independent photometric testing laboratories and provided by reflector manufacturers. Ballast factors are based on lamp manufacturers' data and accepted industry averages. Other light loss factors are obtained from the IES Lighting Handbook, 6th Edition, and from reliable industry sources.

Current typical costs for reflectors, lamps, luminaires, and labor were obtained from reflector and lighting distributors.

TYPES OF REFLECTORS

Reflectors are available in two basic types: semi-rigid reflectors which are secured in the fixture by mechanical means, and adhesive films which are applied directly to the interior surfaces of the fixture. Either silver or aluminum may be used as the reflecting media.

Silver reflectors are made by coating or impregnating a polyester film with elemental silver. The film may be bonded to an aluminum substrate to produce a semi-rigid silvered reflector, or coated with adhesive and applied directly to the fixture.

Semi-rigid aluminum reflectors are made from highly specular anodized aluminum sheet. Aluminized film reflectors are similar to silver films, but have lower reflectance.

Silver reflectors typically have reflectances in the 90% to 97% range, while aluminum reflectors range in reflectance from 70% to 90%. For comparison, white enamel paints used in most older fixtures and many current fixtures typically have reflectances of about 80% to 85%. Modern white powder paints generally exhibit reflectances of about 90%.

Reflector Design

Semi-rigid reflectors are bent to some specific configuration to direct light at desired angles, Figure 9-1.

The shape will be determined by the fixture construction, lamp placement, and desired lighting effect. Some reflectors

Figure 9-1. Two of the many shapes used for reflectors.
Simple shape on left reflects light downward.
Complex shape on right creates multiple lamp
images and has a wider light distribution.

are designed to maximize the amount of light which falls on the workplane by directing most of the light nearly straight downward. These designs may produce high lighting levels directly below the fixture, but areas in between fixtures may be somewhat darker as a result. Other designs direct some light outward at higher angles to provide better uniformity, and to lighten up wall surfaces. The quality of illumination produced by reflectors with wider beamspreads will normally be higher than more concentrating designs; however, the wider distribution units typically exhibit lower efficiencies. Dark walls and areas of low illuminance on the workplane may still be created if fixture spacing is at or near maximum for the unreflectored fixture.

The design of the reflector is critical to performance. Analyses of photometric reports indicate that a properly configured reflector can be expected to improve fixture efficiency by 5% to 15% over a comparable two-lamp fixture without a reflector, or 17% to 27% when compared to a standard four-lamp fixture.

A poorly designed reflector can actually reduce efficiency to less than the efficiency of the fixture without a reflector.

Given comparable good design, a silver reflector can be expected to outperform an aluminum product by about 10%. If the silver reflector is poorly designed, a well-designed aluminum reflector will provide superior performance.

Film reflectors which are applied directly to the interior surfaces of the fixture are generally less efficient than semi-rigid reflectors since they conform to the fixture contours and cannot be formed to direct light in any specific manner.

WHAT HAPPENS
WHEN REFLECTORS ARE INSTALLED

The installation of reflectors in four-lamp troffers is generally accompanied by the removal of all four existing lamps, the installation of two new standard 40 watt lamps, and cleaning of the lens. Each of these actions has an effect on the performance of the fixture, and will occur independent of the installation of the reflector.

The Reflector

Reflectors increase the percentage of lamp lumens which reach the workplane. This is accomplished by redirecting light which is normally emitted at high angles, and the fact that reflectors generally reflect more light than a painted surface. The redirected light travels more nearly downward, so more light travels directly to the workplane, and less light strikes the walls. As previously stated, the redirection of light may result in darkened walls and areas of low illuminance between fixtures.

Effects of Lamp Removal

The lumen output of individual lamps and the percentage of lamp lumens exiting the fixture will both increase when two lamps are removed from a four-lamp fixture. These effects will occur due to a reduction in temperature within the fixtures, and the removal of mass which interferes with the exitance of light from the fixture. They occur simply due to lamp removal, and are not the result of a reflector installation.

Thermal effects result from the operation of fluorescent lamps at other than design temperatures. Standard four-foot lamps are designed to produce rated light output at bulbwall temperatures of about 100°F, which occurs in a still-air ambient temperature of 77°F for standard lamps and about 85°F for reduced wattage energy-saving lamps. Operation at higher temperatures will result in decreased light output, as shown in Figure 9-2a[1]. When bulbwall temperatures are reduced, light output also increases, as does wattage comsumed by the lamps, as shown in Figure 9-2b[2].

Bulbwall temperatures in four-lamp enclosed troffers are difficult to predict, but research[3, 4] indicates that temperatures of 120°F to 130°F can be expected. This results in a decrease in light output of 6.5% to 14%. The removal of two lamps can be expected to reduce bulbwall temperatures by 16°F to 19°F, and the loss of light due to temperature will drop to 0.5% to 2.5%, resulting in an increase in light output of 6% to 11.5%.

RELATIVE LIGHT OUTPUT VS
BULBWALL TEMPERATURE

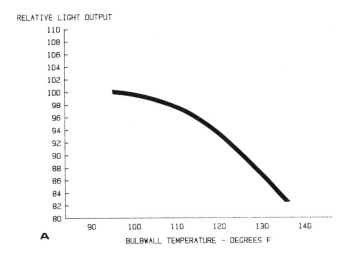

A

F40 SYSTEM WATTS VS
BULBWALL TEMPERATURE

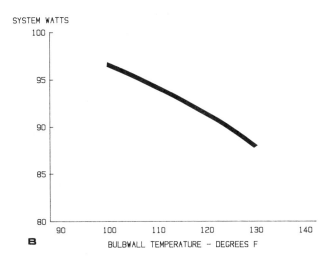

B

Figure 9-2
Thermal effects on light output and wattage consumed by standard F40
lamps. Graph A shows light output as a function of bulbwall temperature.
Graph B shows watts consumed by 2 lamps and standard ballast with
respect to temperature.

Replacement of
Energy-Saving Lamps with Standard Lamps

Energy-saving lamps have been installed in many four-lamp fixtures as an energy conservation measure. When standard lamps are used as replacements, the actual percentage of rated lumens which each lamp produces will increase since most ballasts are designed to operate standard lamps. Energy-saving lamps have slightly different electrical characteristics and, as a result, are not driven as efficiently by the ballast.

In general, standard lamps can be expected to produce about 94% rated light output, and energy-saving lamps about 87% rated light output when operated on commercially available ballasts. When energy-saving lamps are replaced by standard lamps, light output of individual lamps can be expected to increase by about 21% when replacing conventional phosphored lamps, and about 16% when replacing lite white lamps.

This means that two standard lamps will produce about 60% of the total lumens produced by four energy-saving lamps using conventional phosphors, and about 58% of the light obtained from four lite white lamps, simply because the standard lamps are driven more efficiently by the ballast. The percentages may be slightly lower under actual field conditions due to thermal factors.

Replacement of Old Lamps with New Lamps

As fluorescent lamps age they slowly decrease in light output. This is due to a gradual deterioration of the phosphors, and the deposition of evaporated cathode material on the bulb wall. At end of rated life, a standard phosphor lamp will produce about 76% of its initial lumens. Replacing an old lamp with a new lamp will recover this loss.

A typical lumen maintenance curve[5] for 40-watt rapid-start lamps is shown in Figure 9-3.

Effect of Cleaning the Fixture

The gradual accumulation of dirt on luminaire surfaces can have a profound effect on the performance of a lighting system. Lighting systems are seldom cleaned on a regular basis, yet

LUMEN MAINTENANCE
F40 CW. WW. W. LW

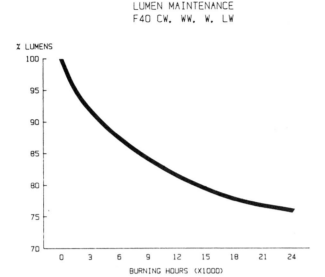

Figure 9-3. Lumen output of F40 lamps as a function of accumulated burning hours.

routine maintenance can be one of the most cost effective ways of reducing lighting costs.

Reflectors are frequently installed in fixtures which have been in service for 5 to 10 years, yet have never been washed. Even in a clean office environment, the loss of light output due to dirt build-up in an unmaintained fixture can be as much as 35%, as shown in Figure 9-4.[6] Simply washing the fixture will recover this loss unless the fixture surface has deterioriate due to lack of cleaning.

When reflectors are installed, the lens is generally cleaned. The reflector is new and clean, and covers the dirty surfaces of the fixture. These actions recover the loss of light due to dirt accumulation on the fixture. The losses may also be recovered through fixture washing, either totally if the reflective properties of the fixture have not been impaired due to lack of maintenance, or at least partially if some deterioration has occurred. Permanent deterioration of luminaire surfaces is difficult to predict, and field evaluation or laboratory testing may be necessary to determine the extent of the problem. Reflectors, or even new fixtures, may be indicated if substantial deterioration has occurred.

Some reflector manufacturers claim that dirt does not adhere to reflector surfaces, and that reflectors are not subject to the same light loss factors as painted fixtures. Supporting data for this claim are not currently available. It appears reasonable, however, to assume that any reduction in light loss will be of small magnitude. An enclosed troffer has four surfaces which may gather dirt: the exterior surface of the lens; the interior surface of the lens; the lamp surface; and the interior surface of the fixture. The distribution of loss over these surfaces is not known; however, only one of the four surfaces is covered by a reflector. It can also be argued that dirt entering the fixture will settle on some surface; and if it does not settle on the reflector, it will probably settle on the lens.

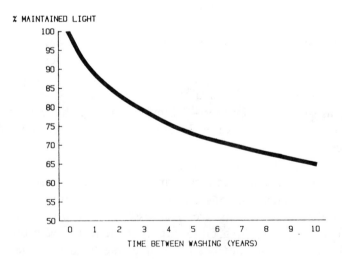

Figure 9-4. Expected loss of light due to dirt build-up for enclosed troffers in a clean office environment.

Multiple Lamp Imaging

Reflectors can create multiple lamp images; a lamp may appear to exist where, in fact, it does not. Multiple imaging creates the visual appearance of having four or more lamps in a fixture which in reality has only two lamps. In some cases this feature may be the sole justification for installing reflectors instead of simply delamping. If aesthetics are a primary consideration, and the appearance of delamped fixture is not acceptable, reflectors may provide the solution.

Alternatives to Consider

When considering reflectors there are sufficient alternatives to fill a small book if each alternative is to be thoroughly analyzed. In addition to the options of various reflectors or simple delamping, there are high efficiency lamps and lamp/ballast combinations which might offer more attractive solutions. For example, if the performance of a delamped fixture without a reflector is marginally acceptable, but slightly more light is desired, high lumen output lamps might provide the answer at a considerably lower cost than reflectors.

The purpose of this chapter is to analyze reflectors and simple delamping; therefore, the other alternatives will not be discussed in detail. They may, however, provide a more viable means of reducing lighting cost, and should be considered.

There are seven basic options which should be evaluated when considering delamping in conjunction with the installation of reflectors in four-lamp troffers:

1. Install silver semi-rigid reflectors.
2. Install aluminum semi-rigid reflectors.
3. Install silver reflective film.
4. Install aluminum reflective film.
5. Delamp without a reflector.
6. Install new two-lamp fixtures.
7. Do nothing, and continue with the existing system.

PERFORMANCE ANALYSIS OF ALTERNATIVES

The average illuminance produced by a lighting system may be predicted from:

$$FC = \frac{(\# \text{ lamps/luminaire}) \text{ (Lumens/lamp) (CU) (LLF)}}{\text{Area per luminaire}}$$

Where: FC = footcandle
        CU   = coefficient of utilization
        LLF = total light loss factor

If photometric data are available for the options which are under consideration, a site specific analysis may be prepared using this method.

The following general case analyses may be useful for preliminary evaluation of the alternatives, provided the parameters resemble conditions at the site. For purposes of analysis, the base assumptions are:

1. The installation is in a 10-year-old office building which operates 12 hours per day, 5½ days per week. The Room Cavity Ratio is 1.0, and the room surface reflectances are 80% ceiling, 50% wall, and 20% floor.

2. The system was originally designed for 100 footcandles (maintained) using standard F40CW lamps with fixtures on 8' x 8' spacing for the first analysis. The second analysis is based on an alternate layout of fixtures on 6' x 8' spacing which produces about 125 footcandles maintained.

3. Dirt condition is clean.

4. Ballast factor is 0.94 for standard lamps and 0.87 for energy-saving lamps.

5. Fixtures will not be washed and lamps will be replaced only upon burnout for the "Not Maintained" analysis.

6. Fixtures will be washed every 12 months and group relamped every 3 years for the "Maintained" analysis.

Eight different scenarios, each with two different maintenance schedules, are analyzed for initial and annual end-of-year illuminance over a 10-year time span (Figures 9-5 and 9-6). Note that the performance of adhesive film reflectors is not included due to a lack of reliable photometric data. It is assumed that films will provide performance somewhere in between semi-rigid reflectors and no reflectors.

The column headed "No-Maint." assumes that luminaires are not washed and lamps are replaced as they burn out. The column headed "Maint." assumes that luminaires are washed on an annual basis and lamps are group replaced every 3 years.

Contition 1 represents the base case, an installation consisting of four-lamp fixtures with standard lamps.

Condition 2 uses the same fixtures but assumes that energy-saving lamps are used. This case is typical of many existing buildings where energy-saving lamps have been retrofitted as an energy conservation measure.

Condition 3 represents the most efficient silver reflector for which photometric data are available. A concentrating light distribution is used to direct light downward to maximize coefficients of utilization.

Condition 4 is the least efficient silver reflector found, and is used for comparison. Most silver reflectors will perform somewhere in between conditions 3 and 4.

Condition 5 is similar to condition 3 except an aluminum reflector is used. Reliable photometric data are not available for this configuration and a coefficient of utilization of 0.83 is assumed, based on a comparison of other photometric tests of aluminum and silver reflectors. While not precise, the estimated CU is believed to be sufficiently accurate for purposes of comparison.

Condition 6 represents the least efficient aluminum reflector for which photometric data are available. As with silver products, most aluminum reflectors can be expected to perform somewhere in between the extremes of conditions 5 and 6.

Condition 7 represents a system using typical new unmodified two-lamp fixtures. The performance characteristics of this system are similar to the aluminum reflector in condition 6, and the average illuminance produced will also be similar.

Figure 9-5 Performance Analysis

End of Year Footcandles
Typical 2' x 4' Troffer in Clean Office Environment
Luminaire Spacing: 8' x 8' Grid Pattern 64 sq. ft/fixt.
Ballast Factor: 0.94 STD Lamps, 0.87 Energy-Saving Lamps

	Condition 1		Condition 2		Condition 3		Condition 4	
	4 STD CW Lamps No Reflector CU = 0.69		4 E/S CW Lamps No Reflector CU = 0.73		2 STD White Lamps Silver Reflector #1 CU = 0.91		2 STD White Lamps Silver Reflector #2 CU = 0.76	
	Footcandles		Footcandles		Footcandles		Footcandles	
Year	No-Maint.	Maint.	No-Maint.	Maint.	No-Maint.	Maint.	No-Maint.	Maint.
0	128	128	111	111	84	84	70	70
1	103	103	90	90	68	68	57	57
2	92	98	80	85	61	64	51	54
3	86	94	75	82	57	62	47	52
4	81	103	70	90	53	68	44	57
5	77	98	67	85	51	64	43	54
6	75	94	66	82	50	62	42	52
7	73	103	64	90	48	68	40	57
8	71	98	62	85	47	64	39	54
9	69	94	60	82	46	62	38	52
10	68	103	59	90	45	68	37	57

	Condition 5		Condition 6		Condition 7		Condition 8	
	2 STD CW Lamps Alum. Reflector #1 CU = 0.83		2 E/S CW Lamps Alum. Reflector #2 CU = 0.73		2 STD WH Lamps New 2 Lamp Fixture CU = 0.73		2 STD CW Lamps Deteriorated Fixture CU = 0.60	
	Footcandles		Footcandles		Footcandles		Footcandles	
Year	No-Maint.	Maint.	No-Maint.	Maint.	No-Maint.	Maint.	No-Maint.	Maint.
0	77	77	68	68	68	68	64	64
1	62	62	55	55	55	55	52	52
2	55	59	49	52	49	52	46	49
3	52	57	45	50	45	50	43	47
4	48	62	43	55	43	55	40	52
5	47	59	41	52	41	52	39	49
6	45	57	40	50	40	50	38	47
7	44	62	39	55	39	55	37	52
8	43	59	38	52	38	52	36	49
9	42	57	37	50	37	50	35	47
10	41	62	36	55	36	55	34	52

Figure 9-6. Performance Analysis

End of Year Footcandles
Typical 2' x 4' Troffer in Clean Office Environment
Luminaire Spacing: 6' x 8' Grid Pattern 48 sq. ft/fixt.
Ballast Factor: 0.94 STD Lamps, 0.87 Energy-Saving Lamps

	Condition 1		Condition 2		Condition 3		Condition 4	
	4 STD CW Lamps No Reflector CU = 0.69		4 E/S CW Lamps No Reflector CU = 0.73		2 STD White Lamps Silver Reflector #1 CU = 0.91		2 STD White Lamps Silver Reflector #2 CU = 0.76	
	Footcandles		Footcandles		Footcandles		Footcandles	
Year	No-Maint.	Maint.	No-Maint.	Maint.	No-Maint.	Maint.	No-Maint.	Maint.
0	170	170	148	148	112	112	94	94
1	138	138	120	120	91	91	76	76
2	123	130	107	113	81	86	68	72
3	114	126	100	110	75	83	63	69
4	108	138	94	120	71	91	59	76
5	103	130	90	113	68	86	57	72
6	101	126	87	110	66	83	55	69
7	98	138	85	120	64	91	54	76
8	95	130	83	113	63	86	52	72
9	92	126	80	110	61	83	51	69
10	91	138	79	120	60	91	50	76

	Condition 5		Condition 6		Condition 7		Condition 8	
	2 STD CW Lamps Alum. Reflector #1 CU = 0.83		2 STD CW Lamps Alum. Reflector #2 CU = 0.73		2 STD CW Lamps New 2 Lamp Fixture CU = 0.73		2 STD CW Lamps Deteriorated Fixture CU = 0.60	
	Footcandles		Footcandles		Footcandles		Footcandles	
Year	No-Maint.	Maint.	No-Maint.	Maint.	No-Maint.	Maint.	No-Maint.	Maint.
0	102	102	90	90	90	90	85	85
1	83	83	73	73	73	73	52	79
2	74	78	65	69	65	69	46	65
3	69	76	61	67	61	67	43	63
4	65	83	57	73	57	73	40	69
5	62	78	55	69	55	69	39	65
6	60	76	53	67	53	67	38	63
7	59	83	52	73	52	73	37	69
8	57	78	50	69	50	69	36	65
9	55	76	49	67	49	67	35	63
10	55	83	48	73	48	73	45	69

Condition 8 represents a typical 10-year-old fixture which has been cleaned and delamped to two standard lamps. Failure to wash the fixture has resulted in permanent deterioration of reflective surfaces and a reduction in fixture efficiency. Note, however, that luminaire surface deterioration can vary widely between installations, depending upon the degree and severity of the contamination which caused the deterioration. This example assumes a reduction in efficiency of about 5%.

Uniformity of Illumination

The previous predictions of illuminance are average illuminance throughout the space. While they provide useful information, they tell nothing about the uniformity of illumination throughout the room, and undesirable areas of light and dark may exist. Uniformity is evaluated by measuring or calculating the illuminance at a series of points in the room. These points are usually on a grid pattern of sufficiently small scale to permit evaluation of changes in illuminance at points under and to the sides of luminaire locations, and at possible work station locations.

Figure 9-7 shows the results of these calculations for a series of points surrounding a single luminaire, as illustrated, in a room measuring 32' x 32', for the four luminaires used in the analysis of average illuminance. A luminaire spacing of 8' x 8' is used. Illuminance is calculated on a 1' x 1' grid pattern, and is approximately the same if transferred to other luminaires within the room. Note, however, that the illuminance in areas adjacent to walls will be slightly lower. These data are applicable only to the luminaires under study, and will vary if luminaires with different light distribution characteristics are used.

Figure 9-8 is similar to Figure 9-6 except that a luminaire spacing of 6' x 8' is assumed.

ECONOMIC ANALYSIS OF ALTERNATIVES

The following economic analyses compare silver semi-rigid reflectors, aluminum semi-rigid reflectors, silver adhesive film reflectors, new two-lamp fixtures, and cleaned existing fixtures

87	89	91	94	94	94	92	90	89
88	90	93	95	96	96	94	91	90
91	93	96	99	100	99	97	94	93
95	96	100	103	104	103	101	98	96
96	98	102	105	105	105	102	99	98
94	96	100	103	104	103	101	97	96
91	92	96	98	99	99	96	94	93
87	89	92	94	95	94	92	90	89
85	87	90	92	93	92	90	88	87

A

54	55	59	63	64	63	60	56	55
54	56	60	64	66	65	61	57	55
56	58	63	69	71	69	64	59	57
57	60	66	73	76	73	67	61	59
58	61	67	74	78	75	68	62	59
57	60	66	73	75	73	66	61	58
55	58	63	68	70	68	63	58	56
54	55	60	64	65	64	60	56	55
53	54	58	62	63	62	58	55	54

B

C

D

49	51	55	58	60	59	55	51	50
50	52	56	60	61	60	56	52	51
52	54	58	63	65	63	59	54	53
54	56	61	66	68	66	61	56	54
54	56	62	67	70	68	62	57	55
53	55	61	66	68	66	61	56	54
51	53	58	63	65	63	58	54	52
49	51	55	59	61	59	56	52	50
48	50	54	58	59	58	54	51	49

48	48	49	50	50	50	50	49	49
48	49	50	51	51	51	50	50	49
50	51	52	53	53	53	52	51	51
52	53	54	55	55	55	54	53	53
52	53	55	56	56	56	55	54	54
52	52	54	55	55	55	54	53	53
50	50	51	52	53	53	52	51	51
48	48	49	50	50	50	50	49	49
47	47	48	49	49	49	49	48	48

Figure 9-7.
Illuminance at a series of points under and around a fixture for 8′ x 8′ luminaire spacing. The numbers in boxes represent the illuminance at points on the grid pattern within the room for several of the options. "A" is a typical unmodified four-lamp lensed troffer, "B" is a concentrating silver reflector, "C" represents a wider spread multiple imaging silver reflector, and "D" is an unmodified two-lamp fixture.

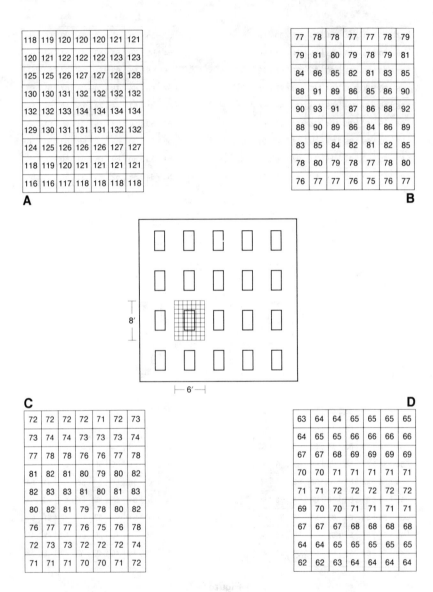

118	119	120	120	120	121	121
120	121	122	122	122	123	123
125	125	126	127	127	128	128
130	130	131	132	132	132	132
132	132	133	134	134	134	134
129	130	131	131	131	132	132
124	125	126	126	126	127	127
118	119	120	121	121	121	121
116	116	117	118	118	118	118

A

77	78	78	77	77	78	79
79	81	80	79	78	79	81
84	86	85	82	81	83	85
88	91	89	86	85	86	90
90	93	91	87	86	88	92
88	90	89	86	84	86	89
83	85	84	82	81	82	85
78	80	79	78	77	78	80
76	77	77	76	75	76	77

B

C

8′

⊢ 6′ ⊣

D

72	72	72	72	71	72	73
73	74	74	73	73	73	74
77	78	78	76	76	77	78
81	82	81	80	79	80	82
82	83	83	81	80	81	83
80	82	81	79	78	80	82
76	77	77	76	75	76	78
72	73	73	72	72	72	74
71	71	71	70	70	71	72

63	64	64	65	65	65	65
64	65	65	66	66	66	66
67	67	68	69	69	69	69
70	70	71	71	71	71	71
71	71	72	72	72	72	72
69	70	70	71	71	71	71
67	67	67	68	68	68	68
64	64	65	65	65	65	65
62	62	63	64	64	64	64

Figure 9-8.
Similar to Figure 9-6, except 6′ x 8′ luminaire spacing is used.

which have been delamped, to an existing system consisting of unmaintained four-lamp fixtures.

In order to evaluate the economics of maintaining approximately the same end-of-maintenance cycle illuminance for the alternatives, it is assumed that fixtures with silver semi-rigid reflectors will be washed and group relamped at 4-year intervals, fixtures with semi-rigid aluminum reflectors will be washed and group relamped at 3-year intervals, and fixtures with silver adhesive film reflectors will be washed every 18 months and relamped every 3 years.

The "new 2-lamp fixture" and "delamped existing fixture" options assume annual washing and 3-year lamp replacement.

Two different scenarios are presented; the first comparison (Figure 9-9) assumes that standard lamps are used in all systems. The second scenario (Figure 9-10) is based on energy-saving lamps in the existing system, and standard lamps in each of the alternatives.

The analyses assume a 20-year life for the systems, and cost of capital at 10%. These parameters will vary, but are assumed to be sufficiently accurate for a general case analysis.

Energy cost savings are addressed only from a lighting standpoint, and additional savings or costs attributable to air conditioning or heating are not included.

DISCUSSION

When considering any major lighting system modification, it must be remembered that the primary objective of the system is to provide illumination of adequate quantity and quality for the task being performed, at an acceptable (and frequently the lowest) cost. Aesthetics may also be an important factor, and must be evaluated with respect to the degree of aesthetics desired and the value they may have to the individual.

Before any conclusions can be reached about the suitability of reflectors for a given installation, the required quantity and quality of illuminance must be determined. Most buildings that are candidates for reflectors were built in an era when lighting systems were designed to produce 100 footcandles or

Figure 9-9. Economic Analysis

BASIS OF ANALYSIS

This comparison evaluates 6 lighting systems on the basis of footcandles and cost. The lighting calculations are made in accordance with procedures established by the Illuminating Engineering Society. The cost evaluations are made using modern principles of life-cycle-cost benefit-analysis. Footcandles produced are the estimated minimum over a 10-year period, with the specified maintenance interval and assuming 8' x 8' luminaire spacing.

The luminaires evaluated in this study are:

Existing: Typical 4-lamp troffer	4 F40CW
No. 1 Existing fixture with silver reflector	2 F40CW
No. 2 Existing fixture with aluminum reflector	2 F40CW
No. 3 Existing fixture with silver adhesive film	2 F40CW
No. 4 Typical new 2-lamp troffer	2 F40CW
No. 5 Existing fixture, cleaned & 2 new lamps	2 F40CW

Bottom Lines	Existing Installation	Proposal No. 1	Proposal No. 2	Proposal No. 3	Proposal No. 4	Proposal No. 5
1. Footcandles produced	68	53	52	N/A	50	47
2. Investment needed	$ 0.00	48280.00	38280.00	31280.00	75280.00	5280.00
3. Capitalized cost	$ 0.00	46000.00	36000.00	29000.00	73000.00	3000.00
4. Total annual life cycle cost	$ 53631.21	32563.53	31678.27	31689.39	37690.95	29468.77
5. Annual savings	$ 0.00	21067.69	21952.94	21941.82	15940.26	24162.44
6. Payback period, years	0.00	2.12	1.66	1.39	3.85	0.23
7. Return on extra investment	0.00	43.64	57.35	70.15	21.17	457.62

Details						
8. No. of luminaires	1000.00	1000.00	1000.00	1000.00	1000.00	1000.00
9. Net cost per luminaire	$ 0.00	0.00	0.00	0.00	43.00	0.00
10. Reflector & labor cost/lum	$ 0.00	46.00	36.00	29.00	30.00	3.00
11. Net cost per lamp	$ 1.14	1.14	1.14	1.14	1.14	1.14
12. Lumens per lamp	3150.00	3150.00	3150.00	3150.00	3150.00	3150.00
13. Annual taxes & owning cost	$ 0.00	5403.15	4228.55	3406.33	8574.56	352.38

	Existing	Proposal 1	Proposal 2	Proposal 3	Proposal 4	Proposal 5
14. Annual power cost @ .08/kWh $	49695.36	25808.64	25808.64	25808.64	25808.64	25808.64
15. kW of load	181.00	94.00	94.00	94.00	94.00	94.00
16. Annual lamp & rep labor cost $	3935.85	811.18	807.75	807.75	807.75	807.75
17. Annual luminaire cleaning cost $	0.00	2500.00	2500.00	2500.00	2500.00	2500.00
18. Coefficient of utilization	0.69	0.91	0.83	N/A	0.73	0.69
19. Lamp lumen depreciation	0.82	0.82	0.84	N/A	0.84	0.84
20. Luminaire dirt depreciation	0.65	0.77	0.80	N/A	0.88	0.88
21. Ballast factor	0.94	0.94	0.94	N/A	0.94	0.94
22. Light loss factor — IES	0.50	0.59	0.63	N/A	0.69	0.69
23. Luminaire washing freq. mo.	100.00	48.00	36.00	18.00	12.00	12.00
24. Relamping interval, mo.	100.00	48.00	36.00	36.00	36.00	36.00

100 months means no luminaire washing or no group-relamping

Figure 9-10. Economic Analysis

BASIS OF ANALYSIS

This comparison evaluates 6 lighting systems on the basis of footcandles and cost. The lighting calculations are made in accordance with procedures established by the Illuminating Engineering Society. The cost evaluations are made using modern principles of life-cycle-cost benefit-analysis. Footcandles produced are the estimated minimum over a 10-year period, with the specified maintenance interval and assuming 8' x 8' luminaire spacing.

The luminaires evaluated in this study are:

Existing: Typical 4-lamp troffer	4 F40CW
No. 1 Existing fixture with silver reflector	2 F40CW
No. 2 Existing fixture with aluminum reflector	2 F40CW
No. 3 Existing fixture with silver adhesive film	2 F40CW
No. 4 Typical new 2-lamp troffer	2 F40CW
No. 5 Existing fixture, cleaned & 2 new lamps	2 F40CW

Bottom Lines	Existing Installation	Proposal No. 1	Proposal No. 2	Proposal No. 3	Proposal No. 4	Proposal No. 5
1. Footcandles produced	59	53	52	N/A	50	47
2. Investment needed $	0.00	48280.00	38280.00	31280.00	75280.00	5280.00
3. Capitalized cost $	0.00	46000.00	36000.00	29000.00	73000.00	3000.00

4.	Total annual life cycle cost	$	47109.52	32563.53	31678.27	31689.39	37690.95	29468.77
5.	Annual savings	$	0.00	14545.99	15431.24	15420.10	9418.57	17640.75
6.	Payback period, years		0.00	2.91	2.28	1.91	5.69	0.32
7.	Return on extra investment		0.00	30.13	40.31	49.30	12.51	334.10
Details								
8.	No. of luminaires		1000.00	1000.00	1000.00	1000.00	1000.00	1000.00
9.	Net cost per luminaire	$	0.00	0.00	0.00	0.00	43.00	0.00
10.	Reflector & labor cost/lum	$	0.00	46.00	36.00	29.00	30.00	3.00
11.	Net cost per lamp	$	1.67	1.14	1.14	1.14	1.14	1.14
12.	Lumens per lamp		2800.00	3150.00	3150.00	3150.00	3150.00	3150.00
13.	Annual taxes & owning cost	$	0.00	5403.15	4228.55	3406.33	8574.56	352.38
14.	Annual power cost @ .08/kWh	$	42831.36	25808.64	25808.64	25808.64	25808.64	25808.64
15.	kW of load		156.00	94.00	94.00	94.00	94.00	94.00
16.	Annual lamp & rep labor cost	$	4278.16	811.18	807.75	807.75	807.75	807.75
17.	Annual luminaire cleaning cost	$	0.00	2500.00	2500.00	2500.00	2500.00	2500.00
18.	Coefficient of utilization		0.73	0.91	0.83	N/A	0.73	0.69
19.	Lamp lumen depreciation		0.82	0.82	0.84	N/A	0.84	0.84
20.	Luminaire dirt depreciation		0.65	0.77	0.80	N/A	0.88	0.88
21.	Ballast factor		0.87	0.94	0.94	N/A	0.94	0.94
22.	Light loss factor – IES		0.46	0.59	0.63	N/A	0.69	0.69
23.	Luminaire washing freq. mo.		100.00	48.00	36.00	18.00	12.00	12.00
24.	Relamping interval, mo.		100.00	48.00	36.00	36.00	36.00	36.00

100 months means no luminaire washing or no group-relamping

more. In 1981, the Illuminating Engineering Society revised the recommended illuminance for most visual tasks, and established a method for selecting the quantity of light required.[7] Most office tasks are listed at 20 to 50 footcandles, with 30 footcandles applying to the typical task and worker. Some tasks, which are more visually demanding, are typically lighted to 75 to 150 footcandles; however, task-oriented lighting is recommended for these higher levels. As a result of the revisions, an average design illuminance of about 50 footcandles is common. Option 5, simple delamping and maintaining an existing four-lamp system spaced on 8' centers presented in the study can be expected to produce over 50 footcandles for almost 2 years after washing and relamping, and drops to 47 footcandles at the end of a 3-year maintenance cycle. If the desired illuminance is 50 footcandles, the system would normally be considered acceptable. For more information on recommended illuminance and the illuminance selection procedure, the IES Lighting Handbook should be consulted.

Once the required illuminance is determined, a site-specific analysis should be prepared. The methodology used in this chapter may serve as a quideline.

For the general case analysis presented it appears that any of the alternatives presented could be acceptable for the typical office, provided that the existing luminaire has not deteriorated appreciably. This can be determined only by testing or a closely monitored field evaluation.

It is common practice to install one or several reflectors on a trial basis to permit field evaluations of performance in the client's facility. This may be a viable method of evaluating; however, the following procedure is recommended:

1. Install new lamps and let them operate for 100 hours to burn in and stabilize.

2. After the burn-in period, remove the lamps and thoroughly wash the fixture, lens, and lamps. Let the lens air dry. Do not dry it with a towel, since a static charge may build up, which will attract dirt.

3. Re-install the lamps and let the system operate until the lamp

bulbwall temperature has stabilized. A period of three hours should be sufficient.

4. Take illuminance measurements at a series of points under and around the fixture. The grid patterns shown in Figures 9-6 or 9-7 are recommended. Make sure that lamps have been removed or power turned off to all other fixtures in the area which might affect the test. Also make sure that no extraneous light from windows or other lights enters the area.

5. Install the reflector and repeat steps 3 and 4. Mark the lamps before removal to assure that they are installed in the same sockets.

When evaluating the results of the test, remember that the measured illuminance is the initial illuminance produced by the system, and will deteriorate over time. If the system is washed annually and relamped at 10,000 burning-hour intervals (about 3 years in most large office buildings), the illuminance at the end of the third year should be about 75% of the initial illuminance.

Also be aware that there are differences between light meters. An inexpensive meter may have an accuracy of 10% or more, while precision photometers typically have 4% to 5% accuracy relating to the U.S. Standard, and 2% to 3% relative error. An inexpensive meter may read 45 or 55 footcandles, when in fact the illuminance is 50 footcandles. Be wary of placing absolute judgment on meter readings unless the characteristics of the meter are known.

Two areas of interest have been omitted from this discussion of reflectors: product durability and cost of washing. Aluminum has been used by the lighting industry for many years and has proven to be acceptable reflector material. Silver films are relatively new and their durability is somewhat unknown. There have been reported instances of film separation and flaking, and some films may have little abrasion resistance. When considering a silver reflector or film, check the manufacturer's track record. Obtain a sample of the material and test its abrasion resistance. Scratches in a specular surface will affect its reflectance characteristics, and result in a more diffused reflectance.

The cost used in the economic analysis for cleaning reflectors is assumed to be the same as an unreflectored fixture since cost data is not currently available. It appears reasonable to assume, however, that reflectors will be more difficult to clean than normal fixture surfaces, thus the cost will be higher.

CONCLUSIONS

From the data presented it appears that, in general, there are three instances where reflectors may be indicated in four-lamp lensed troffers: when the visual appearance of a delamped fixture is unacceptable; when the reflective surfaces of the existing fixture have deteriorated to an unacceptable level due to lack of maintenance; or when the original fixture has an unusually low efficiency, as is typically the case with narrow fixtures, which may be greatly improved by a redesigned reflector.

A comparison of silver and aluminum semi-rigid reflectors indicates a marginal improvement in performance with silver products. Given the higher cost of silver reflectors, it is doubtful that this improvement justifies the added cost. Performance characteristics of adhesive film reflectors are dependent on the fixture configuration; however, it appears that they will be most effective in shallow fixtures.

For the cases analyzed, reflectors may provide an expensive alternative to proper lighting system maintenance. In most cases washing intervals can be increased to three or four years with reflectors, instead of the annual washing which would typically be required if fixtures are simply delamped. The cost of fixture washing, however, is low in comparison to the cost of reflectors, and reflectors require a large initial investment. When the overall economics of reflectors is considered, it is difficult to justify their installation in most applications if simple delamping is aesthetically acceptable.

Most reflectors concentrate light under the fixture, and dark spots or areas of low illuminance may result. The uniformity of illumination should be evaluated to assure that adequate light is provided at workstations, and the uniformity ratios are not excessive.

An evaluation of the quality of illumination for tasks which are subject to veiling reflections, such as typical office work, should include an evaluation of Equivalent Sphere Illumination (ESI) for the various alternatives. The results of ESI calculations for three of the options analyzed in this chapter are shown in the following table:

| | | | Midpoint Between | |
| | Nadir | | Fixt. | Across Dir. |
Reflector	RAW FC	ESI FC*	RAW FC	ESI FC*
Concentrating silver	78	25	58	57
Multi-imaging silver	70	21	54	52
2-lamp fixture no ref.	56	23	53	52

*North-viewing direction
From a qualitative standpoint, the three systems are approximately equal in terms of veiling reflections.

A decision to install reflectors, or any of the other alternatives to reflectors, should be made only after the evaluation of a site-specific analysis, which includes not only economic and quantitative factors, but also a consideration of the qualitative aspects of the illuminance produced by the lighting system. Due to the complexities of interactions between lighting system components, the analysis should be performed only by a qualified lighting professional.

References

[1] Replotted from "Improved 35W Low Energy Lamp-Ballast System," *Journal of the IES,* April 1980, Illuminating Engineering Society, New York, NY
[2] *Ibid.*
[3] *Ibid.*
[4] "Luminaire Retrofit Performance," Lighting Technologies, Inc., *EPRI Project 2418-3,* Electric Power Research Institute, Palo Alto, CA
[5] Replotted from General Electric Company data on lamp lumen depreciation.
[6] Replotted from *IES Lighting Handbook,* 1984 Reference Volume, Illuminating Engineering Society, New York, NY
[7] *IES Lighting Handbook,* 1987 Applications Volume, Illuminating Engineering Society, New York, NY

SECTION IV
BALLAST SELECTION

Chapter 10
Matching Fluorescent Lamps And Solid State Ballasts to Maximize Energy Savings

R.A. Tucker

ABSTRACT

Opportunities are available to maximize energy savings by using advanced energy efficient fluorescent lighting systems. This chapter discusses several systems on the market that benefit from recently introduced technologies.

Increased system efficiencies are possible by utilizing newly developed fluorescent lamps, electronic ballasts, and fixture types with improved optical characteristics. With proper combination of these advanced lighting technologies, an optimum design that provides significant reductions in both initial costs and system operating costs can be attained.

INTRODUCTION

During the past decade, energy consumed by lighting systems has been closely scrutinized for possible cost reduction consideration. For this reason, an assortment of lighting equipment and products were introduced by manufacturers to respond to the demand for increased efficiencies by users. The reason for this interest in lighting is simple. Lighting is important to the building owner because it can represent 50 to 60 percent of the commercial user's electric load. Therefore, from a business point of view, it is vital that we continue to look at lighting energy and, where possible, implement more effective and efficient advanced lighting technologies to cut operating expenses. The high cost of energy and the trend toward compulsory state regulations are focusing added attention on using lighting energy more efficiently.

California, Massachusetts and New York are examples of states with legislation limiting the watts allowed per square foot of space in a new commercial establishment. Legislation has been seen governing the power consumption limits on existing facilities and mandating a timetable to meet these lighting and HVAC limits.

In concert with these legislative actions, many utility companies across the United States offer rebates to commercial users to help fund the installation of energy-saving lamps, energy-saving ballasts, and energy-efficient lighting systems. They are spending large amounts of money to reduce demand and power consumption.

It is apparent that lighting energy management has become a priority. However, it still is an ignored opportunity by a majority of lighting users.

NEW LIGHTING SYSTEM
DESIGN CONSIDERATIONS

Now that designers must operate in an area of heightened energy awareness and regulation, there is much interest in reduced diameter (T-8) fluorescent sources for both task lighting and general lighting applications. Matching these small diameter lamps with an electronic ballast results in higher lamp efficacies (lumens per watt) and increased system efficiencies. With the family of T-8 straight and U-shaped lamps increasing, additional design flexibility exists in terms of available wattages. The low operating current (265ma) for the T-8 family has permitted a lamp design compatible with a 60 Hz rapid start or electronic high frequency (25 KHz) instant start circuitry.

The highly efficient T-8 lamp-electric ballast combination has spurred the imaginations of many luminaire manufacturers as they develop an assortment of new fixtures. The better temperature performance of the T-8 design has allowed greater freedom in fixture design relative to optical and thermal considerations.

Consider the example of a 30-foot by 30-foot office space designed to be lighted by 2-foot by 4-foot parabolic louvered fixtures equipped with magnetic ballast and three 4-foot standard fluorescent lamps. The designer wants to maintain approximately 50 footcandles in this office with a ceiling height of 8.5 feet and the following surface reflectances: 80 percent ceiling, 50 percent walls, and 20 percent floor. Ten deep-cell parabolic 2-foot by 4-foot luminaires will be required representing 1.58 watts per square foot. The more aesthetic deep cell parabolic 2-foot by 2-foot fixture will require 15 units to do the same lighting job, representing 1.53 watts per square feet of lighting load (see Tables 10-1 and 10-2).

If we use an electronic ballast designed for the T-12 energy-saving 34-watt lamps in the 2-foot by 4-foot deep cell unit, then 10 fixtures will be needed to provide the 50 footcandles of illumination for a total of 1.0 watts per square foot. However, three reduced diamter T-8 U-lamps and the T-8 electronic ballast in a deep cell parabolic 2-foot by 2-foot system can outperform most other systems. In this example, the same space can be lighted to the desired illumination with 11 units, requiring only 0.94 watts per square foot. The only system that can outperform the T-8 U-shape lamp system is the T-8 straight lamp with the electronic ballast. It can light the space using 0.84 watts per square foot.

BALLASTS COMPATABILITY

Like all fluorescent lamps, T-8 lamps must be operated on a ballast to limit the current and provide the required starting voltage. The ballast must be designed specifically for the lamp's electrical characteristics, the type of circuit on which it is operated, and the voltage and frequency of the power supply.

Because of its 265 ma. operating current, the T-8 lamp required the design of new rapid start ballasts. The magnetic rapid start ballast used in these lamps offers two primary advantages: (1) economical design and (2) long lamp life of 20,000 hours.

Table 10-1.
System Comparison for Various Lamp-Ballast Combinations

Fixture Type	Lamp Type	Ballast Type	Average Watts/Fixture	Average Watts/Sq. Ft.*
2x4 Parabolic 3-Lamp	T-12 Standard (40W)	Standard Magnetic	142	1.58
2x2 Parabolic 2-Lamp	T-12 U-Lamp (6-in. bend dia.)	Standard Magnetic	92	1.53
2x2 Parabolic 3-Lamp	T-8 U-Lamp	T-8 Magnetic	102	1.25
2x4 Parabolic 3-Lamp	T-12 Energy-Saving (34W)	Electronic	90	1.0
2x4 Parabolic 3-Lamp	T-8 Lamp (32W)	T-8 Magnetic	108	0.96
2x2 Parabolic 3-Lamp	T-8 U-Lamp	T-8 Electronic	77	0.94
2x4 Parabolic 3-Lamp	T-8 Lamp (32W)	T-8 Electronic	84	0.84

*To obtain 50 footcandles illumination in an office with 8.5-foot ceilings measuring 30-foot by 30-foot with surface reflectances of 80 percent ceiling, 50 percent walls and 20 percent floor.

Single and two-lamp magnetic ballasts are currently available for both 120 and 227 volt operation.

High frequency operation (20 KHz and up) improves fluorescent lamp efficiency, affording the opportunity to deliver the same light output for less power. Installations of high-frequency systems have been limited primarily by the cost and efficiency of the equipment required to convert power into higher frequencies.

Instant start circuits prove more economical at high frequency, because circuit design is simplified, with less complex ballasts and fixtures. The cathodes of T-8 lamps were designed with the knowledge that high-frequency operation would become more popular. Specific design considerations were incorporated into the lamps' cathodes to allow them to operate effectively either at high frequency or at 60Hz. Currently there are two-lamp, three-lamp and four-lamp high frequency instant start ballasts available. With these ballasts, lamps operate in parallel, so that if one lamp fails, the others continue to operate. All these high frequency electronic ballasts are also available in 120 and 277 volts.

Table 10-2.
System Comparison for Various U-Lamp-Ballast Combinations

Lamp Type	Ballast	Ballast[1] Factor	Watts[2]	Relative[3] Light Output (RLO)	RLO/W
Standard T-12 U-Lamp 2 Lamp 6-inch bend dia.	Standard Magnetic	.95	92	100	100
Standard T-12 U-Lamp 3 Lamp 3-inch bend dia.	Standard Magnetic	.95	145	128	81
Energy Saving T-12 U-Lamp 2 Lamp 6-inch bend dia.	Energy Saving Magnetic	.88	70	86	113
Energy Saving T-8 U-Lamp 3 Lamp	T-8 Magnetic, 3L	.95	102	138	125
Energy Saving[4] T-8 U-Lamp 3 Lamp	T-8 Electronic, 3L	.92	77	134	160

[1] Data in test normalized to ballast factors shown in this column for magnetic ballasts. Factors shown for electronic ballasts are measured values of sample.
[2] Data were obtained for a system using one-lamp and two-lamp ballasts. 3L designation refers to a three-lamp ballast system.
[3] Relative light output based on initial (100 hour) rated lamp lumen output.
[4] Life rated at 15,000 hours. All other systems shown are rated at 18,000 hours.

While T-8 lamps operated on instant start have life ratings of 15,000 hours, they do provide the advantages mentioned above. The multiple lamp ballast designs simplify the equipment needed not only in four-lamp luminaires (by requiring just one ballast instead of two), but also in the increasingly popular three-lamp systems, which traditionally required both a single and a two-lamp ballast. Due to the increased efficiency of these systems, the shorter lamp life is more than offset by the reduction of power consumed to deliver the same light output. Lumen maintenance is improved to 91 percent at 40 percent of rated life when Octron lamps are operated on the high frequency instant start systems.

RETROFITTING THE EXISTING SYSTEM

An example of the potential for energy savings is the retro-fitting of Tampa University using T-8 lamps and electronic ballasts. It began after extensive testing in a section of the university's library and will eventually extend to some 36 other buildings on campus situated on 69 acres along the Hillsborough River.

The Tampa Electric Company (TECO) initiated the energy reduction program through a monetary grant to the university for energy conservation and offered its expertise to the school. The library was targeted as a test site because its lighting is critical as well as relatively constant.

An area of the library known as the Florida Military Collection Room was selected for the test because the lighting circuits could be isolated for energy use measurement and documentation. Illumination levels recorded prior to the test were an average of 67 footcandles, including the contribution of daylight. The desired illuminance level for the room after relamping was set at a maximum of 100 footcandles. Relamping objectives were threefold: to meet the IES standard at the lowest cost, to provide a high color-rendering quality of light, and to reduce ultraviolet emissions that could harm the valuable paintings and artifacts in the room.

The room contained 42 three-lamp fluorescent fixtures and four four-lamp fixtures equipped with 35-watt cool-white fluorescent lamps. An advantage for the T-8 system was its compatibility with existing bi-pin lamp holders in the 2-foot by 4-foot troffers. After evaluating various lamp/ballast systems (see Tables 11-3 and 11-4) and after meters were in place and a baseline established, 32-watt warm (3100K) T-8 lamps were installed. In addition, 46 electronic ballasts and new lens were installed.

Results immediately after the relamping and refurbishing showed a 38 percent reduction of 1.941 kilowatts, while illuminance levels had increased to 126 footcandles, more light than was needed in the room. Since the ballasts used could operate either three or two lamps, the number of lamps per fixture was reduced to two, for a resulting illuminance of 93.1 footcandles, not considering the contribution of daylight.

Based on the university's energy cost of 7 cents per kilowatt hour and the library's 4,365 hours of annual operation, the Collection Room and an additional 998 lamps in the library are projected to return all costs of the relamping, including new ballasts and maintenance, in less than 27 months.

Table 10-3.
Comparison of Four-Lamp Recessed Troffer Fluorescent Systems

Lamp Type	Ballast	Ballast[1] Factor	Watts	Relative[2] Light Output (RLO)	RLO/W
Standard T-12 , 40 watt	Standard Magnetic	.95	174	100	100
Energy Saving T-12, 34 watt	Energy Saving Magnetic	.88	139	91	114
Energy Saving T-12, 32 watt	Energy Saving Magnetic	.88	131	91	121
Energy Saving[3] T-12, 28 watt	Energy Saving Magnetic	.95	116	89	133
Energy Saving T-12, 34 watt	Electronic	.75	119	91	133
Energy Saving T-8, 32 watt	T-8 Magnetic	.95	032	101	133
Energy Saving[3] T-8, 32 watt	T-8 Electronic	.92	106	98	161

[1] Data in test normalized to ballast factors shown in this column for magnetic ballasts. Factors shown for electronic ballasts are measured.
[2] Relative light output based on initial (100 hour) rated lamp lumen output.
[3] Life rated at 15,000 hours. All other systems shown are rated at 20,000 hours.

Table 10-4.
Comparison of Three-Lamp Parabolic Louvered Fluorescent Systems

Lamp Type	Ballast	Ballast[1] Factor	Watts[2]	Relative[3] Light Output (RLO)	RLO/W[2]
Standard T-12	Standard Magnetic	.95	148 (139)	100	100 (107)
Energy Saving T-12, 34 watt	Energy Saving Magnetic	.88	115 (106)	90	116 (126)
Energy Saving T-12, 34 watt	Energy Saving Magnetic	.88	107 (99)	89	123 (133)
Energy Saving[4] T-12, 28 watt	Energy Saving	.95	102 (93)	86	125 (136)
Energy Saving T-12, 34 watt	Energy Saving Magnetic, 3L	.90	106	91	126
Energy Saving T-12, 32 watt	Energy Saving Magnetic, 3L	.90	95	87	134
Energy Saving T-12, 34 watt	Electronic, 3L	.88	90	90	148
Energy Saving T-8, 32 watt	T-8 Magnetic	.95	108 (104)	97	133 (138)
Energy Saving[4] T-8, 32 watt	T-8 Electronic, 3L	.93	84	96	170

[1] Data in test normalized to ballast factors shown in this column for magnetic ballasts. Factors shown for electronic ballasts are measured values of sample.

[2] Data were obtained for a system using one-lamp and two-lamp ballasts. Data in ( ) are for a tandem wired system using two-lamp ballasts. 3L designations refer to a three-lamp ballast system.

[3] Relative light output based on initial (100 hour) rated lamp lumen output.

[4] Life rated at 15,000 hours. All other systems shown are rated at 20,000 hours.

Chapter 11
Application of Solid State Ballasts

M.S. Gould

INTRODUCTION

Manufacturers' claims of solid-state ballast performance have been of interest to facility engineers for some time. Since these devices are relatively new to the market, little field data on performance exists. As a preliminary step to the installation of solid-state ballasts in a large-scale project, Stanford conducted independent bench testing of several solid-state ballasts. Parameters such as input power, power factor, relative light output, voltage regulation, flicker, harmonic content, and conducted EMI were evaluated in the laboratory. Lamp life and ballast failure are currently being monitored in several small test installations.

The solid-state ballasts tested performed 14% to 20% more efficiently than core and coil ballasts. Percent flicker is 33% with a core and coil ballast and between 2% and 25% with solid-state ballasts. Harmonic content varied from ballast to ballast. These results are discussed in relation to the requirements of particular lighting applications.

BACKGROUND

Visitors are almost always surprised to learn just how energy-intensive a place like Stanford University can be. This is by no means a sleepy college community! Stanford's annual energy bill is over $12,500,000, and lighting represents approximately $2,800,000 of this total. Since energy is an overhead cost, it competes directly for funds with Stanford's academic mission; hence there is a strong desire to control energy expenditures.

The use of electronic ballasts at Stanford must be seen in the context of the University's broader view of such devices. During the early 1980s large-scale delamping projects and incandescent conversions were completed on the Stanford campus. Many of the early energy management projects—such as delamping—were relatively easy to implement, had quick paybacks, and were easy to reverse if necessary. Because energy management remains a worthwhile investment, however, Stanford continually seeks new ways to conserve energy, often using sophisticated, innovative products. Many new companies are formed quickly and introduce new products in an attempt to beat other manufacturers to the marketplace. As a result, products not completely perfected are sometimes introduced. Because of the lack of field experience with these new products, Stanford generally takes a cautious approach.

Nonetheless, electronic ballasts show great promise, and not only as energy-saving devices. They also reduce flicker, a potential problem for those who work around video display terminals (VDT's). There is evidence that the interaction of fluorescent lamp flicker with display refresher rates affects people who work with VDT's. Since electronic ballasts operate in the 20 to 50 kHz range flicker can be significantly reduced.

It is estimated that the lighting costs could be reduced by as much as $250,000 per year if all standard core and coil ballasts on campus were replaced with electronic ballasts. Many buildings are approaching the age where reballasting is required. In addition, a major campus redevelopment effort that is now under way, with at least one large building planned for each of the next 5 years, holds further potential for savings. For these reasons the impact of an energy-saving device such as the electronic ballast could be very large.

Ballasts are required to perform in different environments and fixture types. It is unlikely that one particular ballast will be optimal for every application. Because of this, we are not looking for one ballast to fit all uses, but rather several ballasts that can be applied where they fit best. The goal is eventually to have enough confidence to specify electronic ballasts for all new construction, remodels, and replacements.

Several considerations kept us from a quick entry into large ballast replacement projects. First, we were concerned that high frequency electronic ballasts might interfere with sensitive electronic equipment. Many buildings on campus have computer or lab testing equipment that is costly and delicate. Futhermore, it is imperative that this equipment operate uninterrupted. Second, standard core and coil ballasts are a stable and reliable product. Their technology is straightforward and has been refined over the last several decades. Universally accepted data to compare specific electronic ballasts, however, are not available, although the American National Standards Institute (ANSI) is currently working on a revised version of Standard C82 to include electronic ballasts. Manufacturers' data can be quite helpful in explaining product features, but it is difficult to judge performance with this information alone. Direct comparisons become more difficult when tests are done under different operating conditions. Each manufacturer uses different lamp fixtures and test equipment to measure performance.

Our final area of concern was Stanford's own acceptance of a new technology. Building occupants are justifiably protective of their work environment and suspicious of changes. Likewise, shop personnel are reluctant to risk repeated trips to a project site when they know a core and coil ballast will work fine. Products, like people, earn reputations that can be long lasting. With sufficient research and investigation a good track record can be established and acceptance will follow. The failure of one project might make it difficult to implement future projects.

PROGRAM OBJECTIVES

- Collection and Organization of Literature
- Bench Testing
- Field Testing

The purpose of the bench tests is not to uncover unique data but rather to satisfy ourselves 1) that the electronic ballasts will save energy while maintaining acceptable light output; 2) that the ballasts will maintain this performance through reasonable voltage variations; and 3) that flicker can be improved.

We felt the best way to satisfy our concerns was to do our own independent evaluation of electronic ballasts. The first part of this review was collecting and organizing information on electronic ballasts. Letters were sent out to seven manufacturers requesting product literature, technical reports, and in-house testing results. We also asked each manufacturer to contact our office. Some of this information was very useful. For example, we found that some manufacturers recommend using certain lamps. Independent sources such as the Illumination Engineering Society handbooks were consulted to determine the appropriate procedures and parameters for ballast testing. The Lawrence Berkeley Laboratory Lighting Systems Research Group has also written numerous technical articles on electronic ballast performance. Useful references are listed at the end of this chapter.

After reviewing the above sources, we decided on a bench test procedure that could compare the operating costs and performance capabilities of the ballasts. The next step was to obtain test equipment and find a suitable location for our tests. The tests are not complex and can be done by anyone familiar with basic electric circuits and test equipment such as oscilloscopes and ammeters. There are several companies that rent equipment if purchasing or borrowing is not practical. Equipment required would depend of course on the tests chosen. (See the Appendix for the model numbers of the test equipment used.)

Each parameter that we tested has an impact on either direct lamp operation cost, user comfort, or compatibility with other equipment as shown in the table below.

Testing	Area of Impact		
Parameter	Operation Cost	User Comfort	Compatibility
Input voltage	x		
Input line current	x		
Power factor	x		
Light output		x	
Voltage regulation		x	
Flicker factor		x	
System efficacy	x	x	
Harmonics			x
Conducted EMI			x

System efficacy and the voltage-output regulation parameters were derived, and the other parameters were measured directly.

DATA COLLECTION

Our interest was to compare only the relative performance of these electronic and core and coil ballasts. Absolute measurements would require additional equipment to calibrate the lamp test fixture. (This is usually accomplished in an integrating sphere.) The disadvantage of relative measurements is that they cannot be readily compared with measurements from other sources, but they were satisfactory for our purposes.

A comparison test requires a reference lamp and ballast. The Sylvania GTE Super Saver II LiteWhite F40/LW/RS/SS 34-watt lamp was used with the GE Maxi-Miser II (core and coil) ballast. This combination was selected because it is the most common on campus. Stanford has made an effort to standardize on one type of four-foot fluorescent. We found that stocking both warm whites and cool whites in 40-watt and 34-watt tubes became too unwieldy for our relampers. Burnt-out lamps will most likely be replaced by whatever is in the relamping truck. Therefore, we selected only those ballasts that did not require a special lamp to operate.

Our carpenter's shop constructed a wooden box to cover the lamps during the tests, to simulate an enclosed fixture with a stable operating temperature. The lamp temperature should be held relatively constant since temperature directly affects light output. To speed testing, a small wooden test board was also built with holes for lead connections. This made it easy to plug in the ballast for each test, rather than hand wire each time.

Prior to the active testing, four fluorescent lamps were burned in for 120 hours each. This was done to ensure stable operation during the experiments. Much of the initial lumen depreciation common in fluorescent lamps takes place in the first 100 hours. Lamps were also warmed up for at least 15 minutes before each ballast was tested. Two different ballasts from each manufacturer were used for each test. The ballast testing results are summarized in the Appendix.

TEST EQUIPMENT AND
ANALYSIS OF TEST RESULTS

Input Voltage — This was measured with a digital power analyzer and held constant with a Variac. (A Variac is a variable transformer used to vary and control AC voltage.)

Input Line Current — The power analyzer was also used for measuring the input current in amperes.

Power — Since power consumption directly determines operating costs, it is best to use a power analyzer/wattmeter which will measure true RMS power. Power can also be calculated using the current, voltage, and either a measured or estimated power factor, but this method is not sufficiently accurate.

Power Factor — Power factor can be calculated when the voltage, current, and input power is known (P/VI = Power Factor). Most utilities charge customers for having a power factor under 85%. A high power factor is desirable in order to keep currents down and hence enable the use of smaller-sized conductors. Although performance varied from a power factor of 97.9% to 90.6%, we considered all values to be within an acceptable range.

Lumen Output — Lumen output was tested by taking foot-candle measurements with a high quality hand-held digital light meter. First we measured our reference ballast's light output and compared that with the manufacturer's rated average output for that particular model lamp. From this information we determined a correction factor for the rest of the readings. This method, of course, gives only relative measurements, but is adequate for our purposes. Absolute measurements were beyond the scope of our test equipment.

Voltage Output Regulation — Voltage was adjusted with the Variac to 10 percent above (132 v) and 10 percent below (108 v) standard 120 service. Light readings were then taken to determine how each ballast performed under these conditions. Because small voltage variations can be expected from time to time, the ballast should be able to handle these with a minimum of change in light output. A large change in light output will disturb the building occupants. We also wanted to determine if there was anything in the circuitry to cause them to fail under these conditions. Our tests showed that electronic ballasts appear to be more sensitive to voltage fluctuations than core type ballasts. However, this additional sensitivity does not appear to be significant enough to cause user discomfort. In no case did the ballasts fail to perform in the specified voltage range.

Percent Flicker — As mentioned earlier, one of the important selling points of an electronic ballast is the reduced flicker over standard core and coil ballasts. The *IES 1984 Reference Manual* describes flicker as a cyclic variation in instantaneous light output. Since this variation is twice the input frequency (120 Hz modulation when 60 Hz power is supplied), flicker can be virtually eliminated at the frequency range that electronic ballasts operate in (between 20 and 30 kHz). Using a Tektronix light probe with a small amplifier and an oscilloscope, a single cycle light variation was read and the maximum and minimum values recorded. The formula for percent flicker is then 100 (max−min)/(max + min). Although percent flicker varied from ballast to ballast, the results were lower than core and coil ballasts in all cases.

System Efficacy — Efficacy, in lumens per watt, is the best cost performance measure of the lamp and ballast system performance. Electronic ballasts were 14% to 20% more efficacious than our reference core and coil ballast. In all cases the electronic ballasts used less power (watts) than the reference core and coil. In some cases part of this efficacy improvement is at the expense of light output. This can be a problem where light levels are close to or below minimum recommended levels.

Harmonics and Conducted Electromagnetic Energy (EMI) — Harmonic content can be measured using a spectrum analyzer. In simple terms, harmonics are high frequency variations of the fundamental 60 Hz sinusoidal wave form. The electronic switching in the ballasts can produce harmonics that are transmitted back into the power supply. If there is enough disturbance in the power line, other sensitive equipment (such as personal computers) could be disrupted. In a three-phase power system the neutral can become overloaded with a large third harmonic. EMI effects are potentially higher with the electronic ballasts. However, in some cases they were the same or lower than our reference core and coil. Therefore, EMI readings should only be of concern where sensitive equipment is operated at the same fundamental frequency as the electronic ballast (20-30 kHz).

FIELD TESTING

Field testing is the last part of ballast evaluation. There are several key parameters that cannot be bench analyzed. One is the effect that electronic ballasts have on fluorescent lamp life. The best way to test this is to install the ballasts into typical work environments and monitor their performance. An hour meter can be installed at the circuit panel to measure lamp life. Another concern is audible noise problems. In one field test, electronic ballasts were installed with a manual dimmer. Several ballasts had a noticeable hum that increased as the light level decreased. Some people seem to be more sensitive to this than others. The problem was solved by installing a different brand of ballast.

In one of the field tests we found that additional labor costs for installing new lamp ballast combinations should be considered. For example, some ballast manufacturers offer three-lamp and four-lamp models. Though these can lead to a more efficient lamp ballast system, our relampers were not familiar with these products, and it took repeated trips to the site before they were acquainted with the new wiring requirements. In some cases this doubled the installation time for a normal two-lamp or one-lamp ballast (also part of three-lamp and four-lamp fixtures). Probably this time will be reduced once the relamper has installed a number of units, but it may be a continuing concern.

An interesting side benefit of electronic ballasts came to our attention recently while completing a lighting project in one of our music buildings. For many years the Dinkelspiel Auditorium recital hall had been lit by 22 300-watt incandescent lamps.

When it became necessary to renovate the room and install a drop ceiling it was appropriate to install fluorescent fixtures. The musicians were pleased with the renovation aesthetically but claimed that the hum from the fluorescent ballasts was annoying. When visiting this site it was difficult for the untrained ear to determine the origin of this complaint. Nothing appeared out of the ordinary. When the musicians insisted that something be done we called the Lighting Systems Research group at Lawrence Berkeley Labs to see if they had had similar experiences. As it turns out, a standard core and coil ballast operates at a frequency between the musical notes B and B-flat. Because of this, the musicians using this room had difficulty tuning their instruments. Electronic ballasts were then installed, and because of their higher frequency operation they alleviated the problem.

FURTHER CONSIDERATIONS

Other considerations not yet discussed include ballast cost and manufacturer reliability. Costs are continually changing and can vary with quantities purchased and vendor competition. When contacting vendors we found large variances in cost, availability, and delivery time. As with any large purchase, recommendations from other users should be solicited.

Dimming capability is a feature available with some electronic ballasts. We did test one of the electronic ballasts that had an external dimming feature. Input power, power factor, and voltage regulation tests were repeated at one-third and two-thirds of 100% light output. As expected, power input and light output are nearly proportional. At lower light levels this relationship appears to fall off. This is apparently due to the power requirements of the lamp filament to preserve lamp life. Power factor decreased by 7% to below 85% at one-third light output. Besides the additional energy savings, dimming offers a convenient method of adjusting light levels without the expensive control hardware required for dimming core and coil ballasts. Dimmable ballasts can also work in combination with daylight design strategies. Ballast performance under dimmed conditions is worth further investigation. Our dimming tests were done with only one manufacturer, who has since gone out of business. For this reason the results of the dimming tests are not included in the Appendix Summary.

CONCLUSIONS

- Electronic ballasts result in energy savings at the expense of some light output relative to core and coil ballasts.

- Efficacy of electronic ballast systems is 14% to 20% higher than core and coil ballast systems.

- Electronic ballast systems appear to be more sensitive to voltage fluctuations than core ballast systems. However, the observed variations are not considered significant.

- EMI effects, as expected, were generally more pronounced in electronic ballasts. However, except for environments where equipment is operated in certain frequency ranges (coinciding with 20-30 kHz fundamental), EMI effects were secondary.

- No one ballast exhibited across-the-board superiority in performance (with respect to measured parameters).

- Electronic ballasts performed well in the power factor tests and in all cases had flicker levels below core and coil ballasts.

ACKNOWLEDGEMENT

Our energy management group is in the unique position of being able to hire highly qualified part-time help from our student body. Thomas Ikuenobe, a part-time employee of Stanford's Energy Management Group and a doctoral candidate in the school of Electrical Engineering, conducted the bench tests for us.

Table 11-1. Ballast Test Summary

Ballast Model	# of Lamps	Voltage (Volt)	Current (Amp)	Power (Watt)	Efficacy (Lum/Watt)	Output @120V (Lumen)	Regulation @108V	@132V
Electronic "A"	2	120	0.51	55.8	82.3	4592	-9.69	6.99
Electronic "B"	4	120	0.94	106.7	81.5	8700	-5.31	-0.83
Electronic "C"	3	120	0.76	85.2	79.4	6764	-6.61	9.36
Electronic "D"	2	120	0.48	53.3	78.6	4187	-9.03	7.88
Electronic "E"	2	120	0.50	56.5	78.1	4414	-8.02	8.84
Electronic "F"	2	120	0.46	54.0	77.5	4183	-9.71	11.04
Electronic "G"	2	120	0.49	54.7	75.7	4139	-9.23	8.48
Core and Coil "1"	2	120	0.48	55.9	71.5	3996	-4.38	3.88
Core and Coil "2"	2	120	0.63	68.5	65.5	4490	-4.99	3.01
Core and Coil "3"	2	120	0.62	72.8	65.1	4740	-1.60	2.62

Ballast Model	Flicker (%)	PF (%)	Harmonic as Percentage of the fundamental			Conducted EMI (dbuV)
			(3rd)	(5th)	(7th)	
Electronic "A"	3	90.6	25.0	3.2	1.3	93
Electronic "B"	2	94.7	6.3	6.3	4.0	118
Electronic "C"	8	93.5	32.0	1.0	0.8	113
Electronic "D"	11	93.3	16.0	3.2	1.0	113
Electronic "E"	10	93.6	32.0	1.0	0.8	110
Electronic "F"	26	97.2	10.0	2.0	1.0	115
Electronic "G"	5	93.2	25.0	2.0	1.0	93
Core and Coil "1"	33	97.9	16.6	4.0	0.3	55
Core and Coil "2"	33	90.8	16.0	4.5	0.0	55
Core and Coil "3"	33	97.8	16.0	5.0	0.1	55

EQUIPMENT LIST

1. Two 2-lamp Fluorescent fixtures
2. Variac (0-230V)
3. Fluke 77 Multimeter
4. Fluke Digital Thermometer 2176A
5. Valhalla Scientific Digital Power Analyzer 2101
6. Litemate III Photometer (Photo Research)
7. Tektronix Light Probe J6502
8. Power Supply 0-40VDC 0-5A (Holtek Electronics HR40-5B)
9. Hitachi Oscilloscope
10. Beckmann Industrial Oscilloscope Granit Mate 9020 (20 mhz)
11. HP 3580A Spectrum Analyzer 0-300khz
12. Triplett Current Transformer
13. Polaroid CR-9 Land Camera
14. Wooden Lamp Fixture

References

American National Standards Institute, Fluorescent Lamp and Ballast Committee. C82.1, 1977; C82.2, 1972.

A.A. Arthur, R.R. Verderber, F. Rubinstein, and O. Morse. "Electromagnetic Interference Measurements of Fluorescent Lamps Operated With Solid State Ballasts," *IEEE Transactions on Industry Applications*, vol. 1A-18, No. 6, November-December, 1982, pp. 647-652.

M.G. Harms, L.P. Leung, and R.R. Verderber. *Electromagnetic Interference From Fluorescent Lighting Operated With Solid State Ballasts in Various Sites*, Lawrence Berkeley Laboratory, LBL Report 17998, June 1984

Illuminating Engineering Society of North America, *IES Lighting Handbook*, 1981.

J.E. Jewel, S. Selkowitz, and R.R. Verderber. "Solid State Ballasts Prove to be Energy Savers," *Lighting Design and Application*, vol. 10, no. 10, January 1980, pp. 36-42.

C.R. Stevens and D.W. Aitken. "Energy Savings with Increased Productivity in Today's Office: The Union of Emerging EMS and Controllable Output Ballast Technologies," *Proceedings of the 7th World Energy Engineering Congress*, Association of Energy Engineers, Atlanta, Georgia, November 1984.

R.R. Verderber. "Electronic Ballast Improves Efficiency," *Electrical Consultant*, Cleworth Publishing Co., Inc., November-December 1980, pp. 22-26.

R.R. Verderber and O. Morse. *Performance of Electronic Ballast and Other New Lighting Equipment*, Lawrence Berkeley Laboratory, LBL Report 20119, Oct. 1985.

SECTION V
MICROCOMPUTER SOFTWARE

Chapter 12
A Spreadsheet Template
For Simple Lighting Designs
Using Lotus® 1-2-3®

John L. Fetters, C.E.M.

This chapter describes how Lotus 1-2-3 can be used for simple lighting designs.

LDW: THE LIGHTING DESIGN WORKSHEETS

In this program, the filename on the disk storing the template is LDW and the file contains three sections called worksheets. The template was made user friendly with a menu driven command selection so it only requires that the user have a little working knowledge of Lotus 1-2-3 commands. Lotus is loaded in drive A. The disk with LDW is placed in drive B and the command "/FR" is typed to retrieve the LDW file or any other file that the user has stored on the data disk.

A title screen is presented to the user with instructions (Figure 12-1), along with a menu displayed at the top of the screen for the user to select a worksheet. Selecting the "LUMINAIRES" worksheet allows the user to determine the number of luminaires required to light a given space with a certain light source. Selecting the "FOOTCANDLES" worksheet provides the user with an illuminance level for a given space using a certain number of luminaires and a given light source. Selecting the "GRID-LAYOUT" assists the user with simple rectangular grid layouts. Other menu selections are provided to "PRINT" the work-

1-2-3® and Lotus® are U.S. registered trademarks of Lotus Development Corp.

Figure 12-1

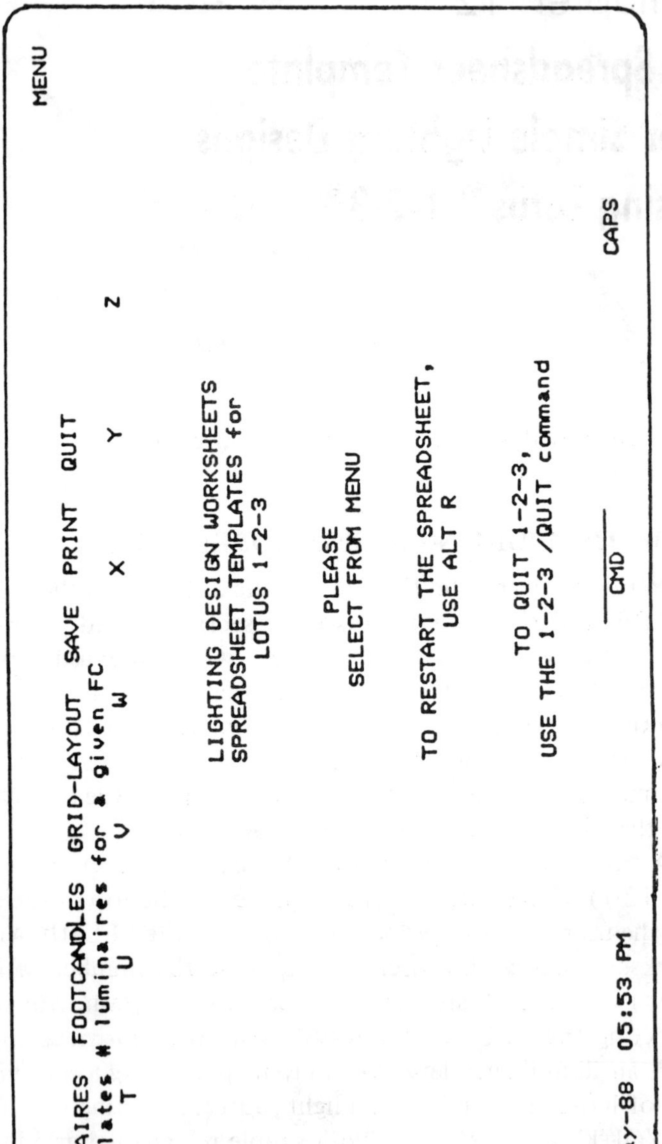

```
T1:
LUMINAIRES  FOOTCANDLES  GRID-LAYOUT  SAVE  PRINT  QUIT                    MENU
Calculates # luminaires for a given FC
       T                      U        V     W      X      Y    Z

 1
 2
 3                                    LIGHTING DESIGN WORKSHEETS
 4                                    SPREADSHEET TEMPLATES for
 5                                         LOTUS 1-2-3
 6
 7
 8                                         PLEASE
 9                                    SELECT FROM MENU
10
11
12                                    TO RESTART THE SPREADSHEET,
13                                         USE ALT R
14
15
16                                         TO QUIT 1-2-3,
17                                    USE THE 1-2-3 /QUIT command
18
19
20
04-May-88   05:53 PM                          CMD                          CAPS
```

sheets, "SAVE" the files on the data disk and "QUIT" when finished.

After the user selects a worksheet, a new menu will be displayed at the top of the screen. The user is asked to "ERASE OLD WORKSHEET" or "MODIFY EXISTING WORKSHEET" (Figure 12-2). This provides an easy method to retrieve and modify an existing worksheet without re-entering all the data again. The spreadsheet will automatically date and time stamp the current document and the user is prompted to enter a name. This name will appear on the worksheet near the bottom of the space identified as "WORKSHEET PREPARED BY:". The program then moves the cursor to each data cell required for the calculations.

A convention has been used to help identify those cells used for inputting data from those cells that operate on data cells to calculate a result. Note that all data input cells are marked by a ":" at the right end of the label, while those calculated cells are identified with a "=".

To help the user identify their lighting project, a two-line space is provided at the cells to the right of the label "PRO-JECT IDENTIFICATION:". Next, the cursor moves to a cell labeled "LENGTH:" where the user is expected to enter the dimensions of the room length. After each data entry, the user strokes the enter key. Most PC's now mark this key with a 90-degree bent arrow. (Some call it a carriage return; a throw-back to ancient typewriters, no doubt!) At each enter stroke, the program moves the cursor to the next data entry cell of the worksheet.

At this point it is assumed that the user understands how to enter the data required of the template where the cursor has guided them. But what if an incorrect entry is made? With these worksheets, it is easier to make all corrections after all data is entered.

After all data for the worksheet is entered, the user may "disconnect" from the automatic cursor advancement by choosing "QUIT" from the command menu. Then the user simply moves to the cell requiring correction using the arrow keys, enters the correct data, and resumes the design process.

Figure 12-2

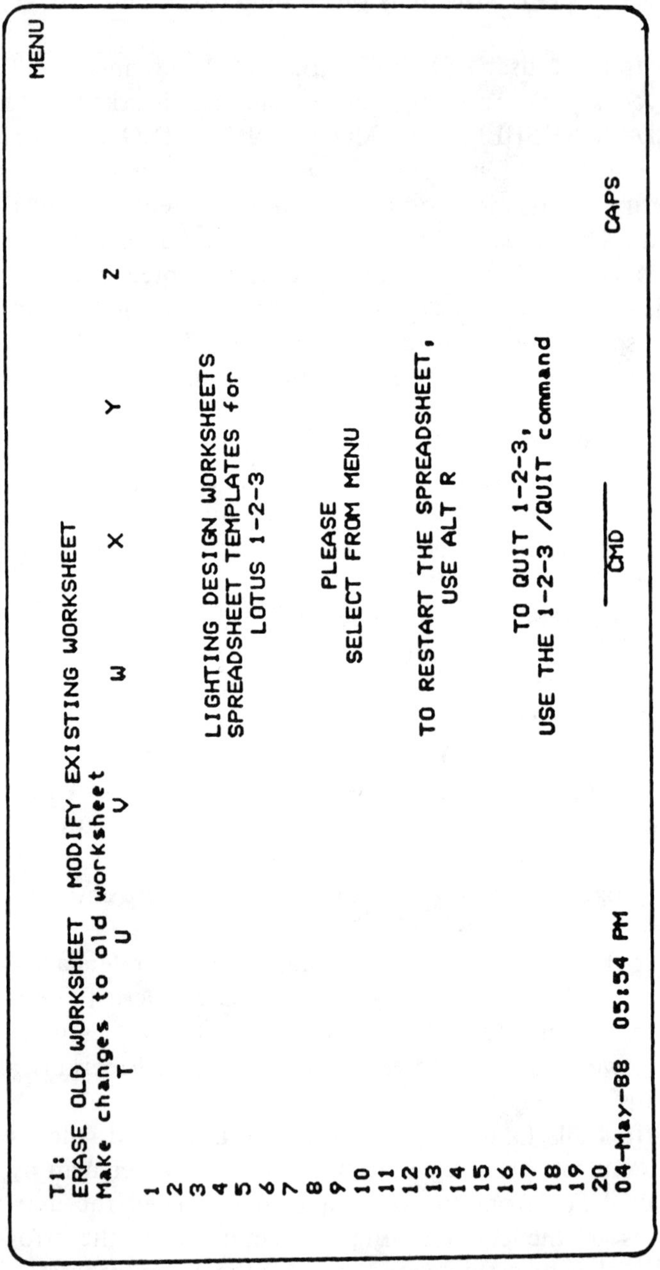

```
T1:                                                                    MENU
ERASE OLD WORKSHEET   MODIFY EXISTING WORKSHEET
Make changes to old worksheet
       T        U        V        W        X        Y        Z
 1
 2
 3
 4                    LIGHTING DESIGN WORKSHEETS
 5                    SPREADSHEET TEMPLATES for
 6                          LOTUS 1-2-3
 7
 8                            PLEASE
 9                       SELECT FROM MENU
10
11
12                   TO RESTART THE SPREADSHEET,
13                            USE ALT R
14
15
16                        TO QUIT 1-2-3,
17               USE THE 1-2-3 /QUIT command
18
19
20
04-May-88  05:54 PM                   CMD                              CAPS
```

THE "LUMINAIRES" WORKSHEET (Figure 12-3)

The "LUMINAIRES" worksheet is used when the designer wants to determine how many luminaires are required to develop an average illuminance in a space. The basis of this worksheet is the IESNA zonal cavity method (also called the Lumen method) and requires that the user provide coefficient of utilization (CU) values obtained from the luminaire manufacturer's catalog. It is intended for simple, direct lighting applications.

After the user has entered values for the room dimensions, the worksheet calculates a value for the Room Cavity Ratio (RCR) and enters this number in the cell labeled "Room Cavity Ratio (RCR)=".

Reflectance values are entered into the worksheet but are not used for computation. These values are recorded only as a handy place to record them as a reminder of what values were used to look up a CU (coefficient of utilization) value in the luminaire catalog, using the computed RCR and the reflectance values. A light loss factor is estimated, based on considerations found in the IES Handbook. Using the light loss factor, the worksheet will calculate the "INITIAL=" illuminance in the cell directly below the cell where the user entered the "DESIGN ILLUMINANCE (AVG):". In the newest version of the template, a separate cell is dedicated to "BALLAST FACTOR:" to show the effect of this variable on the initial light level.

Lamp data, obtained from the lamp supplier's catalog, is entered. "LAMP WATTS:" and "BALLAST WATTS:" values are used in the power calculations; the "UPD=", the "WATTS/LUMINAIRE=", the "LUMENS/WATT=", and the "TOTAL LOAD=" calculated cells. This information is provided to help the designer determine if the lighting design will come in under the power budget set at the "UPD LIMIT:" cell. If the user does not wish to invoke the UPD calculation, they should enter an @NA in the "UPD LIMIT:" cell.

Next, the user, while consulting the luminaire catalog, enters data for "LUMINAIRE MANUFACTURER:", "CATALOG #:", the "COEFFICIENT OF UTILIZATION (CU):", and the "SPACING CRITERIA:". Spacing criteria inputs are used to

Figure 12-3

LUMINAIRES
LIGHTING DESIGN WORKSHEET
USING IES ZONAL CAVITY METHOD
PROJECT IDENTIFICATION:PART 5   SAVED AS:P5DWR2.WK1
DISHWASHER ROOM

ROOM DIMENSIONS LENGTH: 45.0 FEET
                  WIDTH: 19.3 FEET
        MOUNTING HEIGHT: 9.0 FEET
HEIGHT (FIXTURE TO TASK): 6.5 FEET

REFLECTANCES
FLOOR: 0.20
CEILING: 0.70
WALL: 0.50

ROOM CAVITY RATIO (RCR)= 2.41

LIGHT LOSS FACTOR: 0.80
UPD LIMIT: 1.7 W/SF

DESIGN ILLUMINANCE (AVG): 55` FC
               INITIAL= 66 FC

LAMP TYPE: FLUOR
CATALOG: F40LWSS

LUMINAIRE MANUFACTURER:DAY-BRITE
        CATALOG #:CG142-CO2A

INITIAL LUMENS: 2925
LAMP WATTS: 34
LAMPS/LUMINAIRE: 2

CU (FROM CATALOG): 0.57

BALLAST WATTS: 6.0
BALLAST FACTOR: 0.90
WATTS/LUMINAIRE= 74
LUMENS/WATT= 79 per
              LUMINAIRE

SPACING CRITERIA(L-WISE): 1.20
SPACING CRITERIA(W-WISE): 1.20

MAXIMUM SPACING (L-WISE)= 7.8 FEET
MAXIMUM SPACING (W-WISE)= 7.8 FEET

TOTAL LOAD = 1.5 KW
      UPD = 1.7 W/SF
WHICH IS 100.5% OF
       THE UPD LIMIT

THEOR # OF LUMINAIRES= 20
ACTUAL # OF LUMINAIRES: 20
SQUARE FEET/LUMINAIRE= 43

DATE: 13-May-88
TIME: 05:19 PM

WORKSHEET PREPARED BY:J.L.FETTERS

calculate the maximum spacing values. ("MAXIMUM SPAC-
ING (_-WISE):" When the user enters the CU value, the worksheet calculates a theoretical number of luminaires and enters
it in the cell labeled "THEOR # OF LUMINAIRES=". The
cursor now stops at the cell labeled "ACTUAL # OF LUM-
INAIRES:" so that the user may enter the integer number of
luminaires, rounding or making the number even, if required.
Although it may not be possible for the designer to know the
exact value of this integer on the first pass, a number close to
the theoretical number of luminaires will allow the worksheet
to calculate the "SQUARE FEET/LUMINAIRE=", the "TOTAL
LOAD=", and the "UPD=". Examination of these calculated
values will permit the designer to decide if the design at this
point meets some of the design criteria.

The program prompts the user to choose another type of
worksheet, choose to "PRINT", "SAVE", or "QUIT".

THE "FOOTCANDLES" WORKSHEET (Figure 12-4)

The "FOOTCANDLES" worksheet is used when the designer wants to determine the illuminance of a space for a given
number of luminaires, using a particular luminaire and light
source combination. For example, it would be used to compute
the actual value of illuminance resulting when the actual number of luminaires in the "LUMINAIRES" worksheet is different
than the theoretical number calculated by the worksheet. This
worksheet also uses the IESNA zonal cavity method and, like
the "LUMINAIRES" worksheet, also requires the user to supply
a CU value obtained from the luminaire manufacturer's information. Data values are entered in the same way as described
for the "LUMINAIRES" worksheet. The major difference between the "FOOTCANDLES" worksheet and the "LUMIN-
AIRES" worksheet is that the calculation for illumination and
number of luminaires is reversed, and in this worksheet, an
average footcandle level is calculated from lamp and luminaire data. This makes it convenient for the user to try different
lamp and luminaire combinations to test the sensitivity of the
selected equipment on light levels and power levels.

Figure 12-4

FOOTCANDLES
LIGHTING DESIGN WORKSHEET
USING IES ZONAL CAVITY METHOD
PROJECT IDENTIFICATION:PART 5   SAVED AS:P5DWR2.WK1
                    DISHWASHER ROOM

ROOM DIMENSIONS   LENGTH: 45.0 FEET   REFLECTANCES
                    WIDTH: 19.3 FEET   FLOOR: 0.20
         MOUNTING HEIGHT: 9.0 FEET   CEILING: 0.70
HEIGHT (FIXTURE TO TASK): 6.5 FEET   WALL: 0.50
                    LIGHT LOSS FACTOR: 0.80

ROOM CAVITY RATIO (RCR)= 2.41   UPD LIMIT: 1.7 W/SF

ACTUAL # OF LUMINAIRES: 18   LAMP TYPE: FLUOR
SQUARE FEET/LUMINAIRE=   48   CATALOG: F40LWSS
                                   INITIAL LUMENS: 2925
LUMINAIRE MANUFACTURER:DAY-BRITE   LAMP WATTS: 34
          CATALOG #:CG142-CO2A   LAMPS/LUMINAIRE: 2
                                     BALLAST WATTS: 6.0
CU (FROM CATALOG) 0.57   BALLAST FACTOR: 0.90
                                   WATTS/LUMINAIRE=   74
                                     LUMENS/WATT=   79 per
SPACING CRITERIA(L-WISE): 1.20   LUMINAIRE
SPACING CRITERIA(W-WISE): 1.20

MAXIMUM SPACING (L-WISE)= 7.8 FEET   TOTAL LOAD =   1.3 KW
MAXIMUM SPACING (W-WISE)= 7.8 FEET   UPD =   1.5 W/SF
                                        WHICH IS   90.5% OF
ILLUMINANCE (AVG)= 50 FC   THE UPD LIMIT
          INITIAL= 60 FC   DATE: 13-May-88
                                       TIME: 11:50 AM

WORKSHEET PREPARED BY:J.L.FETTERS

THE "GRID-LAYOUT" WORKSHEET (Figure 12-5)

The IES Zonal Cavity method assumes a uniform layout to achieve a uniform average light level in the space. Many task/ambient lighting designs will require the ambient portion be uniform. Since many of the simpler, fluorescent ambient designs will be accomplished using troffers in a suspended grid ceiling, the Grid-Layout worksheet is used to provide a quick troffer layout.

AN EXAMPLE DESIGN SOLUTION

To illustrate how the LDW template is used, an actual design problem will be shown in detail. The space to be illuminated is a remodeled 45 ft by 19 ft dishwasher room, with a new suspended ceiling at 9 ft. The task at the dishwasher, loading and unloading areas is 30 inches off the floor. The reflectance value of the ceiling tile is .70, the ceramic floor tile is .30, but the effective cavity value due to the equipment in the room is reduced to .20, and the wall reflectance is estimated to be .50.

An adjacent area uses a luminaire that the client wishes to use again to keep the ceiling look the same. The luminaire to be used is a 2-lamp, 1 x 4 fluorescent troffer. The client prefers to use energy-saving lamps, since this is their standard lamp in this facility. To keep their energy-savings program on track, the client demands that the unit power density (UPD) not exceed 1.7 watts/sf. In the old dishwasher room the lighting was described as "dim" and was measured to be 40 footcandles.

In this example, the type of luminaire is known and the solution becomes one of determining how many luminaires are required and where to place them. The grid and room size determine the maximum number of luminaires that will practically fit into the room. The design solution was begun by using the "LUMINAIRES" worksheet first (Figure 12-3), choosing a target footcandle illuminance of 55 fc. After the project description "Part 5 Dishwasher Room" is entered, the room dimensions (LENGTH at 45 ft, WIDTH of 19 ft 3 in), mounting height of 9 ft, and the height of the fixture to the task of (9 ft minus 30 inches) are entered, and the template calculates a

Figure 12-5

LUMINAIRE LAYOUT WORKSHEET
for SPACING IN REGULAR ROWS

PROJECT IDENTIFICATION:PART 5 DISHWASHER ROOM

ROOM LENGTH:45.0
ROOM WIDTH:19.3
ROOM AREA= 866   sq ft

LENGTH of LUMINAIRE: 1.0   feet
WIDTH of LUMINAIRE: 4.0   feet

ROOM LENGTH/LUMINAIRE LENGTH= 45
    LENGTHWISE
    SPACE BETWEEN LUMINAIRES: 6.0
CALCULATED LUMINAIRES per ROW= 6.4
ACTUAL # of LUMINAIRES per ROW: 6 ,leaving 9.0 feet at the ends.

CALCULATED THEORETICAL # of ROWS= 3.3
              ACTUAL # of ROWS: 3

    ACTUAL # of LUMINAIRES= 18 , making 48 s.f./luminaire.
    WIDTHWISE
SPACE BETWEEN LUMINAIRES: 2.0 ,leaving 3.3 feet at the ends.

WORKSHEET PREPARED BY:J.L.FETTERS DATE: 13-MAY-88
                                   TIME: 11:43 AM

Room Cavity Ratio (RCR) of 2.41. The reflectance values (.2, .7, and .5), LLF of .8, UPD LIMIT of 1.7 W/SF, and AVG DESIGN ILLUMINANCE of 55 FC are entered.

Lamp data, found in the lamp manufacturer's catalog and the value of 2 LAMPS/LUMINAIRE are entered next. The Day-Brite catalog number is entered and the catalog data is looked up to find the CU value. Part of the Day-Brite catalog sheet is shown reprinted in Figure 12-7. The photometric data CU table shows that for a 2-lamp luminaire, with an RCR of 2.41 (approximately halfway between RCR = 2 and RCR = 3), and floor reflectance (pfc) of 20% (.20), ceiling reflectance (pw) of 50% (.50), the CU value can be interpolated between 54 and 60 to be 57 (.57). Spacing to mounting height (S/MH) value of 1.2 is entered in place of the spacing criteria. An energy-saving ballast was used that has a loss of 6 watts and a ballast factor of .90. These values are found in the ballast manufacturer's catalog.

The spreadsheet template computes the "Theor # OF LUMINAIRES=" as 20, and 20 is then entered in the cell after "ACTUAL # OF LUMINAIRES:" to provide the template with an integral number of luminaires. When 20 is entered in at "ACTUAL # OF LUMINAIRES" the "UPD" calculates to 1.7 W/SF, "WHICH IS 100.5% OF THE UPD LIMIT". This shows that the maximum number of luminaires used that will not exceed the UPD limit is 20. In addition, the template calculates a "MAXIMUM SPACING" of 7 ft, 9 in, 43 "SQUARE FEET/LUMINAIRE", 74 "WATTS/LUMINAIRE", 79 "LUMENS/WATT per LUMINAIRE", and a "TOTAL LOAD" of 1.5 KW. These calculated values provide the designer with comparative figures for other design iterations.

Since the luminaires are to be installed in a suspended grid ceiling, the "GRID-LAYOUT" worksheet is selected next. The worksheet shown in Figure 12-5 titled "LUMINAIRE LAYOUT WORKSHEET for SPACING IN REGULAR ROWS" comes up on the screen. The room dimensions are copied to this worksheet automatically by the template and the user begins data entry at "LENGTH of LUMINAIRE:" Here the designer must choose how to orient the 1 x 4 luminaires for the tasks to be illuminated. The long dimension of the luminaires is to be parallel with the short dimension of the room, because of equipment layout and to provide good side illumination for the racking operations. This decision requires the entry of 1 foot for "LENGTH of LUMINAIRE:" and 4 feet for "WIDTH of LUMINAIRE:"

Figure 12-6 shows this orientation. Since 2 x 4 panels are to be used between the luminaires an integral number of 2- or 4-foot panels must be considered when choosing the "LENGTH-WISE SPACE BETWEEN LUMINAIRES:" of 6 feet. The worksheet calculates a "CALCULATED LUMINAIRES per ROW=" of 6.4. The user then enters the integer value of 6, since partial luminaires are not allowed! When the value of 6 is entered at the "ACTUAL # of LUMINAIRES per ROW:", the worksheet calculates a value of 9.0 feet "at the ends." This is the total length remaining at both ends of the 45-foot length after 6, 1-foot luminaires have been placed 6 feet apart. The worksheet calculates a value of 3.3 "CALCULATED THEO-RETICAL # of ROWS=", and the user enters the integer value of 3 for "ACTUAL # of ROWS:" 18 "ACTUAL # of LUMIN-AIRES=" is automatically calculated and 48 " s.f./luminaire."

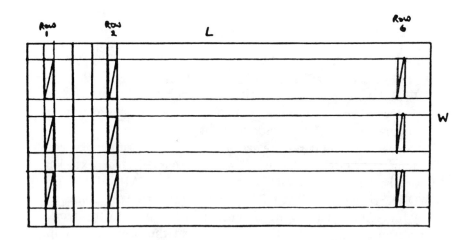

Figure 12-6

The user must now choose a value for "WIDTHWISE SPACE BETWEEN LUMINAIRES:". If the value chosen causes the "leaving __ feet at the ends" value to go negative, the user chooses a smaller spacing, in 2- or 4-foot spacing, until the total length remaining at both ends of the 19-foot, 4-inch width is positive and of an acceptable value. Recall from the "LUMINAIRES" worksheet that a value for "Maximum Spacing" was calculated (Figure 12-3) and the spacings of the luminaires on this Luminaire Layout worksheet must be less than these values for an even distribution of light.

Figure 12-6 shows the final layout of 3 rows of 6 fixtures per row. This results in 18 luminaires which is less than the 20 required to meet the target illuminance of 55 FC. To determine what the actual average illuminance of the space would be using 18 luminaires, the designer chooses the "FOOTCANDLES" worksheet (Figure 12-4). Data entry is the same as described for the "LUMINAIRES" worksheet, except that for the "ACTUAL # OF LUMINAIRES:" 18 is entered, instead of the theoretical value of 20 calculated previously. The worksheet calculates values for "ILLUMINANCE (AVG)=" of 50 FC and for "INITIAL=" of 60 FC. These actual illuminance values are acceptable for this design, so the design process is terminated by printing and saving the three worksheets. If the actual illuminance values were not acceptable, the process would have been repeated until an acceptable layout with acceptable values of illuminance resulted.

SUMMARY

The LDW spreadsheet template has been valuable for simple lighting designs to enable the author to calculate several different solutions for a given lighting problem in a short design interval. Follow-up measurements made after the lighting design has been implemented show good correlation. In some cases where the results were different than expected, the major cause is due to reflectance values which are usually chosen too high for the final finish selection, which was probably not made by the lighting professional.

Of further benefit, a large library of lighting solutions has now been saved, for preferred luminaire and lamp combinations that further cuts the design interval and frees the designer to spend more time with the quality aspects of the design.

References

IES Lighting Handbook, 1984 Reference Volume, John E. Kaufman, PE, FIES, Editor

Figure 12-7.

PHOTOMETRIC DATA

Coefficient of Utilization-Zonal Cavity Method																	
2-Lamp									3-Lamp								
pfc	20								pfc	20							
pcc	80			70			50		pcc	80			70			50	
pw	70	50	30	70	50	30	50	30	pw	70	50	30	70	50	30	50	30
RCR									RCR								
0	76	76	76	74	74	74	71	71	0	67	67	67	66	66	66	63	63
1	71	68	66	69	67	65	64	63	1	63	60	58	61	59	57	57	55
2	66	61	58	64	60	57	58	55	2	58	54	51	57	53	50	51	49
3	61	55	51	59	54	50	52	49	3	54	49	45	52	48	44	46	43
4	56	50	45	55	49	44	47	44	4	50	44	40	49	43	39	42	38
5	52	45	40	51	44	39	43	39	5	46	39	35	45	39	35	38	34
6	48	41	35	47	40	35	39	35	6	43	36	31	42	35	31	34	30
7	45	37	32	44	36	31	35	31	7	39	32	28	39	32	28	31	27
8	41	33	28	40	33	28	32	28	8	36	29	25	36	29	25	28	24
9	39	30	25	37	30	25	29	24	9	34	26	22	33	26	22	25	21
10	35	27	22	35	27	22	26	22	10	31	24	20	31	24	20	23	19

Test No. 7630-2 S/MH=1.2   Test No. 7644 S/MH = 1.2

Average Brightness (Footlamberts) with 3150 Lumen Lamps

2-Lamp		Angle	3-Lamp	
End	Cross		End	Cross
1316	1242	45	1756	1604
950	866	55	1276	1130
743	698	65	996	931
747	698	75	1009	943
702	635	85	922	815

Light Loss Factor Data

LLF	=	0.77	Light Loss Factor (LLF)		
LLF	=	Light Loss Factor		=	LLDxLLLxBF
LDD	=	Luminaire Dirt Depreciation IES Category V Clean Annually	LDD	=	Very Clean 0.93 Clean 0.88 Medium 0.82
LLD	=	Lamp Lumen Depreciation	LLD	=	0.88 40% Rated Lamp Life
BF	=	Ballast Factor (commercial ballast performance relative to reference ballast)	BF	=	0.94 (Std. Ballasts & Lamps) Relamp 70% Lamp Life

Chapter 13
A Microcomputer Program
For Fenestration Design

G.D. Ander, M. Milne, M. Schiler

Simplified computer programs which model a wide variety of fenestration designs in terms of thermal impacts, lighting impacts, and peak load changes are not readily available. This type of design tool is necessary to evaluate fenestration systems in order to make cost-effective design decisions.

While daylighting strategies and the associated photoelectric controls will directly impact electric lighting loads and mechanical loads, the building peak load also changes. Electric utilities have responded to peak load issues by designing incentive programs and time-of-use rates. If one can assume that rational decision makers respond to pricing signals, the associated demand charges can frequently rationalize a daylighting design.

A microcomputer program called DAYLIT has been developed for the building community to address these issues. The program will run on a machine with 256K of memory which makes it compatible with computers of most small to medium size firms. It is extremely user friendly with built-in defaults and help screens. Output can be either tabular or in a graphic form. An hourly simulation program has been developed using published data for both daylighting and heat transfer calculations. It has been packaged so that very little computer literacy is required. It is distributed through Southern California Edison and UCLA. Its use would improve a utility's load factor (peak load/average load).

A COMPUTER ANALYSIS TOOL

The authors believe that for a computer analysis tool to be useful, it needs to be accessible, quick, and reasonably accurate. Feedback from design practitioners and lighting experts during the design of the input screens helped develop this microcomputer program as a tool for making early design decisions. The clear message heard during this review process was that daylighting design concepts could not be justified by knowing the footcandle level at a given point in a room. The bottom line often came down to the cost of a system and the implication on operating costs.

With this in mind, input screens were developed for:

— Cities (climate location)
— Window design
— Skylight design
— Room design
— Lighting system design
— H.V.A.C. system design
— Schedules
— Rates
   (Peak, mid-peak, demand charges, etc.)

It is important to note that the concept of time-of-use rates for electricity and demand charges were identified. Many utilities have these rates and they are designed specifically for decision makers who respond to pricing signals. The cost per kWh and the kW demand charge are extremely high during the "peak period."

Daylighting by virtue of its optimal performance characteristics during these peak hours (usually noon-6 p.m. in the summer) offers a building owner substantial reductions in operating costs. This program is unique in allowing the user to input these rates. By accurately calculating the electric lighting reductions, mechanical reductions and rate structures, the designer can cost justify the fenestration designs of buildings.

DESIGN DATA INPUT

DAYLIT is very carefully developed to be a user friendly design tool. This means that no special training is needed to run this program, since it is possible to immediately begin producing useful output. This is because the interactive dialog is structured so that the user's first intuitive response is usually the correct one. In this case, user friendliness also means that it is not necessary to use the program regularly in order to retain proficiency.

The computer always contains all the necessary data for the design of a real building. Thus, in order to create a totally new design, all the user needs to do is to change one or two key input variables. DAYLIT will immediately calculate and display the daylight curves for the new scheme. In fact, there are eight different data input screens that give the user the opportunity to revise well over 100 design variables (see Figure 13-1). The Window Design Screen is typical. Only 11 design variables are needed to define any window's orientation, glazing, sun controls, light shelves, and overall dimensions. Any term that the user does not understand can be defined by simple typing "help" at the bottom of the screen. Catalogs of glass transmittances can be displayed by simply typing in "single pane" or "double pane" (see Figure 13-2).

The other data input screens include Climate Data, Skylight Design, Room Design, Lighting System Design, HVAC System Design, Schedules, Rates for Electricity and Gas, plus the Library Screen that keeps track of all the user's different schemes and combinations. DAYLIT allows users to input design information, in any order they choose, by simply typing in the name of the next desired screen. If a screen name is forgotten or mistyped, the Options Screen shows a menu of all the 51 possible data input screens, catalogs, graphic output screens, and data tables.

The ultimate proof of this program's commitment to user friendliness is the fact that it can automatically print out a 50-page user's manual on the spot. All the user needs to do is to type in "print users manual."

```
DAYLIT: Daylight Design Tool (For Testing Only 04/86) SCE/UCLA   1/ 1/80  0:20
                                              Project : DEMONSTRATION
   WINDOW DESIGN:                             Climate : LOS ANGELES AIRPORT CZ06
   Scheme 1 BASE CASE


    0.  Window Facing off True South (East = 270. or -90. degrees)
   80.0  % Window Area that is Glass (window minus mullion; 0.0 means NO Window)
   12.0  Head Height FEET (use ceiling height 12.0 or lower)
    3.0  Sill Height FEET (use work surface height or higher)
   70.0  % Glass Transmittance
    6.0* Overhang Width FEET perpendicular to window wall (0.0 means no overhang)
      *  % Diffuse Blind Transmittance (100.% means no blind)
      *  Venetian Blind Angle from horizontal (over 90.Degrees means no blind)
         (venetian blinds are always assumed to block all direct sun)
    0.0  Interior Light Shelf Width FEET (at sill perpendicular to window above)
    0.0  Exterior Light Shelf Width FEET (at sill perpendicular to window above)
   For a catalog of glass transmittances type in SINGLE or DOUBLE PANE below


   Do you want to change any of these values ? (Y,HELP or just hit ENTER)  _
```

**Figure 13-1. The Window Design Screen Shows How Typical
Design Data is Input to DAYLIT**

```
DAYLIT: Daylight Design Tool (For Testing Only 04/86) SCE/UCLA   1/ 1/80  0: 5

   OPTIONS
      Design Screens:              Analysis Screens:
   LIBRARY                      ILLUMINATION LEVEL PLOTS:
   CITIES                          ....month=....(JAN FEB ALL),hours=....(ie.8 to 18)
   CLIMATE DATA                 ANNUAL PERFORMANCE PLOTS: in 3-D
   SINGLE PANE GLASS               ....FOOTCANDLES: at....(MAX MID or MIN)
   DOUBLE PANE GLASS                  ....DAYLIT
   PLASTIC TRANSMITTANCES             ....ELECTRIC LIGHT
   WINDOW DESIGN SUMMARY              ....TOTAL LIGHT
   SKYLIGHT DESIGN SUMMARY         ....ENERGY CONSUMED: in....(KWH BTUH)
   ROOM DESIGN SUMMARY                ....LIGHTING POWER      ....HEATING FUEL
   SURFACE REFLECTANCES               ....EQUIPMENT POWER     ....HVAC ENERGY
   LIGHTING SYSTEM DESIGN             ....AIR CONDITION POWER ....TOTAL ENERGY
   HVAC SYSTEM DESIGN              ....LOADS ON HVAC: in BTUH
   SCHEDULES                          ....LIGHTING COOLING LOAD    ....EQUIPMENT
   RATES                              ....VENTILATION + INFILTRATION ....PEOPLE
   GLOSSARY  (HELP)                   ....GLASS SOLAR + CONDUCTED  ....TOTAL LOAD
   PRINT USER MANUAL               ....COST OF ENERGY : in DOLLARS
   TERMS OF USE                       ....ELECTRICITY....GAS....TOTAL
   MANUAL                       TABLE(S).... (of any of the above)
   WRAP UP (EXIT, STOP)
   Type in any option(s) you want (or HELP or just hit ENTER):  _
```

**Figure 13-2. The Options Screen Appears Whenever Requested
to Show All the Various Design Data Input Screens
and Output Data Analysis Screens**

DAYLIGHTING CALCULATIONS

The basic method used for all of the daylighting calculations was the lumen or flux method, which is the standard Illuminating Engineering Society (IES) method (also known as the L-O-F method). In an area such as Southern California which experiences an extremely large number of clear sky days, this is more accurately handled by keeping the diffuse and direct components separate, which is done in the IES methods. Actual weather patterns are duplicated by working clear and overcast sky conditions separately for each hour of each day, and then combining them in a weighted average based on the ratio of clear to overcast days experienced.

Although curves were determined to match the Coefficient of Utilization tables for sidelighting (C and K values), the tables are stored internally instead with no interpolation to maintain the standard nature of the calculations. Solar position is calculated internally.

The IES methods result in only three data points. Futhermore, the method for calculating overhang values requires interpolation between the first three data points calculated to determine two of the final values (the MAX and the MID values). To graph the values and to be able to interpolate, it was necessary to provide a curve-fitting algorithm. This algorithm defines a curve from the three originally calculated points.

The original IES tables for externally available light were replaced by the new RP–21 algorithms from IES. Otherwise, everything is a duplication of the IES hand calculations except the treatment of light shelves and toplighting.

Toplighting was done using the formulas published in the 1984 IES Lighting Handbook for flat, domed, and double-domed skylight types. The exception is the determination of the values for the efficiency of a light well based on the well wall reflectance and the well which is determined from the height, width, and length (IES Fig. 7-38). The plots reduced to three equations as coded in the program:

for 80% reflectance −]
WELLEFF = e ** (−1.15 * WI),

For 60% reflectance −]
WELLEFF = e ** (−0.89 * WI),

for 40% reflectance
WELLEFF = e ** (−0.55 *WI),

where,
WELLEFF is the well efficiency, and
WI is the well index, calculated from the formula:
WI = (depth * (width + length) / (2 * width * length).

The IES method allows any number of skylights of a single type and size but distributes them evenly throughout the room. The program allows different sizes and types by running them separately and combining them subsequently. The toplighting algorithms are based on the lumen or flux method as used with electric lighting fixtures. The internal reflections and the room shape are done using the Room Cavity Ratio. As a by-product of those assumptions, the light level from toplighting is even throughout the room and a single skylight in a large room is badly modeled.

Light shelves are modeled by breaking the window at the light shelf into upper and lower halves. Each half is modeled separately, and the outside reflectance for the half above the light shelf is adjusted. The effects of the two reflective plans are combined in a weighted average. The effect of the light shelf is based on the viewed angle of the light shelf from the mid-window height at mid-room. The effect of the ground plan is based on the viewed angle of the ground plan from the same position. The glass line is moved inward or outward, as necessary, using the same algorithm as is used for overhangs. There is no theoretical limit to the number of light shelves which can be modeled but practical considerations such as memory and disk storage have limited the number to four.

Clear windows, diffusing blinds, venetian blinds, light shelves, and skylights may be modeled and combined as long as the sidelighting strategies are parallel. For example, opposite walls may contain windows, but adjacent corner walls may not.

ELECTRIC LIGHTING CALCULATIONS

After the daylighting curves are calculated, the electric lights can be "turned on." The architect defines this lighting system in terms of the required footcandle level, and the watts per sq ft (power density) needed to achieve it.

The control system for these lights can be as sophisticated as the user chooses, including up to three zones, each with either continuous dimming or selective switching of any combination of lamps. Control set points can be established for interior zone minimum illumination levels, for safety lighting levels during unoccupied hours, and for lower limits to prevent flickering if fluorescents are dimmed continuously.

Once the electric lighting loads are added to the available daylight in each zone, the electricity consumption for each hour of each month is calculated. From this DAYLIT goes on to calculate the thermal loads from lights.

DAYLIT also contains the option of defining a rather sophisticated time-of-use rate structure. This means it can calculate and display the operating costs of various lighting design schemes. An important and poweful feature unique to DAYLIT is the "compare" command that shows only the differences between two schemes, in this instance, the dollars saved by virtue of a specific design decision.

THERMAL CALCULATIONS

Thermal loads, solar position, and incident radiation are calculated using the standard ASHRAE algorithms. DAYLIT accounts for the heat gain and loss from lights and ballasts, occupants, equipment, infiltration and ventilation, plus radiant and conducted heat flow through glass. For simplicity, conduction through the opaque envelope is ignored, a reasonable assumption because it invariably represents only a tiny percentage of the total loads in modern buildings.

To deal with these thermal loads, DAYLIT allows the user to define different generic heating, ventilating, and air conditioning systems. Economizer cooling and various thermostat set points and dead bands are all available to help reduce energy consumption. DAYLIT calculates hour-by-hour the loads on any of these HVAC systems, the energy they consume, and the cost of that energy.

OUTPUT GRAPHICS

All of the data calculated by DAYLIT can be presented in graphic form. For instance, illumination levels at each foot of the room can be plotted on the cross-section drawings as a series of curves for each hour of each month. Another of DAY-LIT's unique and powerful features is the "combine" command that allows the room to have any combination of windows with overhangs and light shelves, as well as various kinds of skylights. Curves can also be plotted for the total amount of daylight brought in by these combinations of windows and skylights (see Figure 13-3).

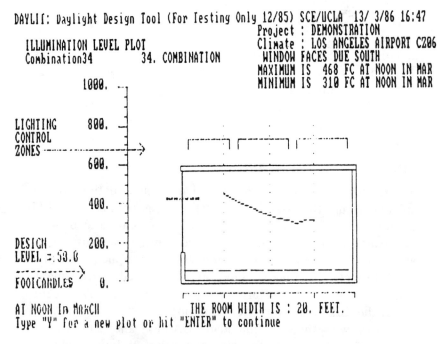

Figure 13-3. Illumination Level Plots Can be Requested for Any
Combination of Hours During the Day and
Months of the Year

The most fascinating graphics produced by DAYLIT are the Annual Performance plots which are displayed in 3-D data for every hour of every month. The Options Screen shows that 16 different performance variables can be plotted in this format beginning with Footcandles of Daylight at the Max Point and ending with Dollars Total Cost of Energy (Figure 13-4).

The data that is plotted out by DAYLIT can also be printed out in tabular form, similar to data tables produced by a spread sheet program.

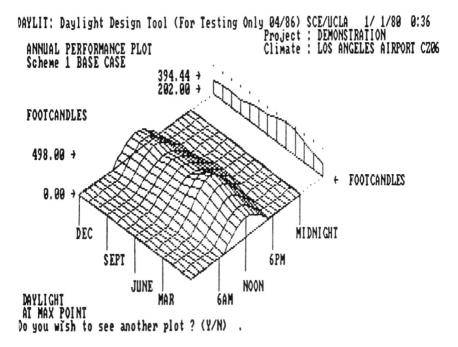

Figure 13-4. This Annual Performance Plot Shows
the Footcandles of Daylight at a Point
5 ft Inside the Window

FUTURE DEVELOPMENTS

Very complex room geometrics are difficult to model, as are sophisticated fenestraton controls. Designs which cannot be directly coded are better analyzed with the physical model.

A data acquisition system has been developed with input screen compatible to the Daylit Program. The next logical phase would be to link the results of the model test directly to the analysis program. This will ultimately give the user unlimited flexibility in quantifying daylighting features.

ACKNOWLEDGEMENTS

The authors would like to recognize the support of the following individuals at Southern California Edison Company—Charles McCarthy, Geoff Bales, Greg Rogers, Brian Brady, Dave Ferguson, Frank Schultz, Emad Hassan, Lily Yoshizumi, Hal Grutbo, Ed Cook, Sharon Hitchcock, Gary Sutliff, and Sandy Deppen, and at UCLA—Cherng-Fong Sheu, Shan-Jean Hwang, Kwok Chan, Den Wun Lin and Rosemary E. Hewley.

References
Ander, Gregg D., et al., "Simplified Daylighting Savings for Nondaylighting Building Energy Simulation Programs," *Buildings and Energy,* 1984 Vol. 6, pp. 221-228.
ASHRAE Handbook—1981 Fundamentals (Atlanta: ASHRAE 1981).
IES Lighting Handbook (New York: Illuminating Engineering Society of North America, 1981).

SECTION VI
NATURAL DAYLIGHTING

Chapter 14
Natural Daylighting — An Energy Analysis

R.P. Jarrell

INTRODUCTION

As energy prices continue to rise, it is becoming imperative that our country's high rate of consumption be reduced. While new technologies are available, and are being developed to ease this consumption rate, significant reductions can be realized through the design and construction of energy efficient structures using existing technologies.

Building owners and developers must realize that they can no longer construct buildings with little or no concern for their energy consumption. If this practice continues, these owners and developers may find that their structures are too expensive to own and operate because energy prices have continued to rise.

Great reductions in a structure's energy consumption rate can be achieved with little change in its appearance or construction and, often, with little increase in the initial construction cost. Higher initial construction costs can be recovered in reduced operating costs.

With these factors in mind, an investigation into one aspect which affects the energy efficiency of a structure is presented. This study investigates the implications of using daylighting in a building and how it impacts the thermal load.

From this entire investigation, the author illustrates how architects have a great potential for providing energy efficient structures. Also highlighted are ways in which the energy performance of a structure can be greatly improved in the initial design phase. Finally, an analysis procedure is developed which will aid architects and others in the design of energy efficient structures.

PROCEDURE:

For this investigation, an existing large-scale office building located in Atlanta, Georgia, was selected. The Georgia State Twin Office Towers project selected was designed in 1977 and was considered an energy efficient structure when completed.

The building's specific design features, construction methods and materials were determined. Daylighting availability levels were established. Other factors such as the building's mechanical and electrical equipment and energy management techniques were determined. Actual building usage data was developed through owner interviews and on-site investigations.

From this data, the energy consumption of the project was estimated. This was accomplished by performing a LOADCAL computer analysis on a representative portion of the building. A typical open plan office level was selected for analysis. The analysis accounted for specific building conditions, existing within the facility, as well as actual environmental factors.

Upon completing this analysis, the availability and usage of daylighting in the existing building was determined. This procedure involved estimating daylighting levels at multiple locations within the typical office level.

Then, alternative building schemes were analyzed to determine if a more efficient design could be provided. This revised scheme did not, however, significantly alter the present architectural design parameters of the facility.

With the objectives and methodology of this report outlined, a detailed description of the Twin Towers will be presented. The discussion will highlight the building's major construction features and materials. Electrical and mechanical systems are described along with energy saving and management techniques utilized in the design, construction and operation of the facility.

PROJECT DESCRIPTION

The Twin Towers office complex consists of four distinct interconnected facilities: the Georgia State MARTA station, the entry plaza, the office towers, and the central energy plant.

The project, located across Martin Luther King Drive from the Capitol, is also bounded by Piedmont Avenue, Butler Street and by the Georgia Railroad trackage.

The MARTA Station is composed of a lower concourse, allowing entry from Piedmont and Butler, an elevated platform where the trains are boarded and an upper concourse which provides entry to the 'plaza' of the Twin Towers. The station is a totally independent facility in terms of operation, maintenance and security. The only link to the Twin Towers is to provide access to the rapid-rail system.

The Plaza is the structure which serves as the base beneath the office towers. Within its four levels are maintenance shops, storage and mechanical space on the lowest level; a cafeteria and snack bar, post office and mechanical space on the second level. The third level is the main public access level and contains office space for state agencies which require maximum public contact. The upper plaza level coincides with the upper concourse of the MARTA station and serves as the access point between the building and the MARTA system.

The two towers rest on the plaza base and each tower contains 16 typical office floors. Each floor is designed to accommodate both open and/or traditional office plan layout, but open office layouts are predominant. (For this study, a typical open plan office floor was selected.) A mechanical penthouse caps each tower.

The fourth facility is the Central Energy Plant. It is partially underground and located adjacent to the south of the towers structure. Provisions for landscaping and access walkways have been made on its roof and thus it sits under the 'front-yard' of the Twin Towers. The central energy plant provides chilled water and electrical service to the Twin Towers.

Brief History:

The impetus for the construction of these facilities was the CAPITOL HILL 2000 plan, a long-term master plan for development of facilities in the vicinity of the state capitol. A key recommendation of the master plan was the construction of two symmetrical state office buildings to house many of the state agencies. The buildings were to be located in air-rights over the proposed rapid-transit station. This would afford both state employees and the general public easy access to these agencies and to Capitol Hill in general. At the same time, the office space provided would enable the state government to increase its efficiency by consolidating, into one location, many of the personnel who were located through the downtown area in leased office space.

In early 1975, the architect, Aeck Associates, Inc., was selected to prepare preliminary studies for the towers and its plaza. This task lasted into early 1976. Simultaneously, MARTA with its own consultants commenced designing the station which would interface with the office tower project. The preliminary drawings were completed for the Twin Towers in February 1976 and the final design phase began. This phase lasted through late 1977. Construction began soon thereafter and continued through November 1981. The building was occupied in phases and became fully occupied in June 1983.

The Building Concept:

As stated, the architect was charged with designing a facility with two identical towers. In order to improve the constructability of each tower, the number of floors was reduced from the concept outlined in the CAPITOL HILL 2000 concept. Furthermore, the area of each floor was increased to improve the space utilization within each tower. This measure eliminated the need for two tiers of elevators, which afforded both economic and energy savings. Because of the basic premise of the planning concept, the two-tower scheme, no consideration was given to combining the towers into a single larger-scaled tower.

Architectural Considerations

The building skin consists of masonry and glass. Prefabricated, medium-color utility brick panels were selected as the predominant exterior cladding material. This selection was based upon a study of aesthetics, construction costs, life-cycle costs and thermal mass.

During the design, studies were made analyzing glazing

types and amounts of glass areas. These studies were based on aesthetics, natural versus artificial light, HVAC loads and views to the outside. The amount of glass utilized in the final design, as a percentage of the building enclosure within a typical 25-foot-wide office bay, is approximately 39 percent. The glass on the north and south elevations is shaded by a brick spandrel which projects 18 inches. East-west facing glass is shaded by a 15-inch brick spandrel. Brick column surrounds, which serve as vertical fins, help to shade the glass with north-south projections of 20 inches and east-west projections of 34 inches.

The glazing selected is Solarban 550-60. It is a one-inch insulated, bronze-tinted glass manufactured by Pittsburgh Plate Glass. The following table describes the design data on the glass used in the Twin Towers.

U-value (summer)   = .59
U-value (winter)   = .50
Shading Coefficient   = .46
Visible Light Transmission = 32%

Table 14-1. Glass Design Data

Various window framing solutions were analyzed including operable windows, provided with and without integral blinds. Operable windows, which appeared a common sense solution, were shown not only to be cost prohibitive (approximately $500,000 in additional costs) but were counter-productive from an energy use standpoint.

By having a fixed window frame, the possible introduction of latent heat from humid summer air is reduced. In winter, the possible loss of heat through exfiltration is reduced by eliminating the large amount of window perimeter 'crack' found with operable windows. Exterior airborne dust intake is also reduced, which has adverse effects upon HVAC equipment.

Several types of insulation are used in the project. All construction U-values met those recommended by the codes in force at the time of the design and construction. The following table gives data on the major types found in the building.

Polyurethane and Perlite
Roof Insulation   U-value = .06

Polystyrene Deck
Insulation   U-value = .06

Rigid Fiberglass Masonry
Panel Insulation   U-value = .11

Table 14-2. Insulation Data

Medium color, horizontal blinds have been provided through-out the entire project. The one-inch blinds have a positive stop at 70 percent of the fully closed position. The blinds reduce the amount of direct solar radiation gain within the spaces. It was pointed out by the mechanical engineer that the blinds reduce the window shading coefficient by, a maximum of, 28 percent to .33.

$$\text{Shading Coefficient} = .46$$
$$.46 \times (.28) = .1288$$
$$.46 - .1288 = .33$$

Mechanical equipment penthouses are located on the rooftop of each tower. The mechanical engineer noted that with a lapse rate of 3½ degrees (F) per 1000 feet, the location of the mechanical equipment on the roof affords the dominant cooling cycle a slight advantage over systems with a lower-level intake.

Mechanical Considerations

The engineer, Nottingham, Brook and Pennington, P.C., commenced by analyzing alternative fuel sources for the building. Coal, solid-waste, oil and natural gas were considered as was off-peak generation. Using the Trane Corporation's TRACE mechanical analysis program, energy analyses were performed evaluating eight options of HVAC equipment and energy sources. These analyses included such items as chiller and condenser equipment and fan equipment on variable volume systems. The building envelope was also analyzed using a computer program based on ASHRAE's Standard 90-75.

HVAC Features Incorporated: A variable-air-volume ventilation system is employed throughout the building. No 'reheat' is allowed within the system. According to the TRACE program, vane-axial supply and return air fans were selected which give the maximum fan power savings and the maximum turn-down ratio available.

Rotary total heat exchangers are used to capture approximately 72 percent of the loss/gain of the central toilet exhaust system. Because of the extended operation time of the building, this affords considerable savings over the course of a year. An air-side economizer cycle, with enthalpy control, reduces mechanical cooling requirements below 78 degrees (F) outdoor temperature. It also provides free cooling below 55 degrees (F) outdoor temperature.

Water wash electronic air filters are employed to provide a lower pressure-drop, thus reducing fan horsepower requirements. The filters selected have an efficiency rating of 95 percent.

Heating is provided by low-temperature, hot-water, finned-piped convectors located under the perimeter windows of the building. This is the most efficient location and method for introducing heat into exterior spaces. The system is zoned by exposure to avoid overheating when solar radiation is available.

Space temperature and ventilation control is provided by induction boxes. These boxes, formally called 'heat-of-light' boxes, reclaim up to 50 percent of the heat from the lights by inducing warm air from the ceiling plenum to the exterior areas where heat is needed in winter. In the summer, this heat is returned to the mechanical penthouse where it is removed by the refrigeration system.

A two-way valve control system is employed on all chilled and heating water coils which provides reduced pump horsepower during lighter loads.

Boiler Plant: Two hot-water generators are incorporated into this facility. They are dual-fuel, 'D' water-tube type with high turn-down ratios (10:1). The water tube boilers in this size (30 million Btu/hr were the most effective ones on the market at the time of construction. Because the boilers are set up on a dual-fuel capability, the owners have the opportunity to take advantage of synthetic fuels as they become economically available. They are presently set up to use natural gas and number two fuel-oil.

The hot water pumping system selected has variable speed capability to reduce pump power consumption during light loads. Heat reclaimed from the hot water pumps' mechanical seal cooling system is used to heat the boiler room and to preheat combustion air.

Control Systems and Automation: A central control room has been incorporated into the Twin Towers and from this point all systems within the towers are monitored. Critical functions are controlled by building engineers from this room.

The owners purchased a Johnson-Controls JC-80 building automation system in an effort to reduce energy consumption and to improve the control of the indoor environment. The towers were designed to incorporate the required conduit, to each floor, to interface mechanical control panels with the JC-80 system.

A number of energy conserving control measures are employed in the operation of the facility. Space temperature setbacks are employed during non-occupancy hours. Both the ventilation and the exhaust systems are automatically shut off during the pre-occupancy warm-up period. Automatic air-temperature and hot-water resets are employed based upon outside air temperature.

Electrical Considerations: Because occupancy requirements were not known prior to construction, considerable flexibility had to be designed into the lighting system. Both open and traditional office layouts were anticipated in the tenant occupancy. Therefore, a variety of conditions had to be addressed in the design of the lighting system.

Given the constraints cited above, four sources of illumination were analyzed using a life-cycle cost comparison. They are mercury, fluorescent, metal-halide and high-pressure sodium. Fluorescent lighting was selected based upon this analysis.

While comparisons of color, appearance and rendering properties were made, cost was the final determinant in selecting a four-tube fluorescent troffer system for general office illumination. The Fluorescent system provides the state's requirement of 70 footcandles, for standard office occupancy, with approximately 2.4 watts per square foot energy usage. This system provided a first-cost saving of about $407,000 over runner-up high-pressure sodium fixtures. The high pressure sodium system would only reduce the annual utility costs by approximately $27,500. The troffers installed are slotted to vent heat directly into the ceiling plenum.

High Intensity Discharge source fixtures have been provided in the lower levels of the plaza where greater ceiling heights exist. They are also used for general outdoor lighting. A minimum of incandescent lighting is used in seldom occupied spaces such as electrical and janitorial closets.

The owner undertook an energy conservation program to reduce the light-wattage, per square foot, in the system provided. In the open plan portions of each tower, two of the four lamps were disconnected in many in reception, waiting and circulation areas. The ballasts serving these delamped fixtures, however, were not disconnected. While the delamping of fixtures reduced the power consumption, greater reductions could be realized by disconnecting the unused ballast in each fixture. A final measure included individual switching of lights in individual offices. These steps were implemented under the direction of the project's design team.

Peak Shaving: Emergency generators for elevators and other electrical equipment are arranged so that they can be brought on-line as needed for electrical demand control. The central control system automatically shuts down selected mechanical and electrical equipment as a power conserving measure and operates the emergency generators to reduce peak electrical demands.

LOADCAL ANALYSIS

In the preceding section, items such as the building's major construction and materials, electrical and mechanical systems and the energy management systems were highlighted. These items were presented and discussed to highlight the systems and components used within the project. With this description provided we can now focus upon the analyses to be performed on the project. These analyses will be the basis for recommending any modification to the structure.

An energy analysis was performed on a representative portion of the building: a typical open plan office level. LOADCAL, the analysis program used, is based on the ASHRAE GRP-158 analysis procedure. It calculates the cooling and heating load of a space (or building). The program takes into account environmental and interior factors which affect the mechanical requirements of the space or building. To ensure that the results of this analysis were as representative as possible, an intermediate office floor was selected. This eliminated any effects caused by having a roof transfer load.

The typical open office floor was divided into four perimeter zones and one interior zone. Each perimeter zone has one exterior exposure. The orientation of the northern zone (zone four) is 40.82 degrees east of true north.

Floor, wall and glazing areas were taken directly from the architectural plans. Wall configurations, overhang and vertical fin dimensions were also derived from the plans. Data concerning U-values were established based on the information contained in the drawings and specifications. They were verified with the project's mechanical engineer.

Typical Atlanta weather data were used in this analysis. The data were taken from meteorological tables, with 1984 being the base year. The solar radiation intensity level, however, was set to equal 1.0 for all evaluations. This value represents a clear sky condition. Actual levels vary, of course, with the degree of cloud cover, level of smog and/or other atmospheric conditions which will block available solar radiation and, thus, reduce the intensity. The effect of using the 1.0 value is a somewhat overestimated solar radiation load than found under normal sky conditions. The increased loading condition used for this analysis will be used in the revised building analysis, thus, allowing for comparable results. Actual intensity data for Atlanta is being developed by others, at this writing.

Analysis data for interior areas represent actual building use conditions. This information was obtained through on-site inspections and from verification with the owners of the project, the Georgia Building Authority. Thermostats maintain a relatively constant year-round temperature of 74 degrees (F).

Humidity ratios used in this analysis are 0.0060 pound/pound for the winter months and 0.0103 pound/pound in the summer months which represents typical Atlanta design data. It is the ratio of the mass of water vapor to the mass of dry air.

Light wattage data were calculated from the architectural and electrical drawings. They were verified through on-site visits and in discussions with the owner. Lighting levels, in footcandles, were checked using a hand-held General Electric illumination meter. The readings indicated that a minimum level of 70 footcandles existed throughout the office areas. Some areas had higher illumination levels.

The light-wattage value established for each zone represents lighting and equipment loads for that zone. Equipment loads were factored by the ballast factor, of the lighting equipment, to eliminate the possibility of over-estimating the total zone wattage value. Light fixture and lamp data were taken directly from the architectural drawings and specifications.

The ballast factor was obtained directly from General Electric, the supplier of the fixtures and the lamps. The delamping plan, as implemented by the owner, was accounted for in the determination of the lighting wattage data. The lighting usage data represent actual conditions, as the lights are left on continually from morning occupancy until the cleaning crews have finished late at night.

The building is typically occupied Monday through Friday, from 8:00 a.m. until approximately 5:00 p.m. The occupant loading is based upon the owner's design criteria of one person per 150 square feet of floor area. This closely represents actual conditions as state government buildings are occupied less densely than typical speculative office buildings. Zones one through four reflect this occupant load condition while zone five is a service zone and the occupant loading here, which is much less, was estimated.

The ventilation rate is ten (10) cubic feet per minute (per person) within the building. The mechanical system employs a rotary total heat exchanger, which is 72 percent efficient, thus the impact of the ventilation rate on the mechanical system is reduced as shown in the following equation.

Ventilation Rate = 10 cfm/person

Heat Exchanger = 72 percent efficient
    (1.0 − .72 = .28)

Energy Consumption from Fresh Air Induction =
    10 * .28 = .28 cfm/person

The system operates only while the building is occupied and therefore a "non-continuous" factor was entered into the LOADCAL program.

LOADCAL estimates the annual cooling and heating load to be approximately 1,307,394,251 Btu's, per year, for the existing office floor analyzed. This translates into approximately 64,088 Btu/sf/year for the typical, 20,400-square-foot, office level.

An appendix which contains the actual LOADCAL computer analysis of the representative portion of the facility will be available for review at the seminar. The input data, found on the initial pages, contain general project data, outside weather data and inside target data. Specific zone-by-zone data for all bays, derived to analyze the building, are also included.

Results from the LOADCAL analysis follow the input data and have been provided in a zone-by-zone format. A floor summary, which sums the results of the zones one through five, is also provided.

This concludes the LOADCAL energy analysis of the existing building. The following section estimates the available daylighting levels. This was accomplished using the methodology developed by Libbey-Owens-Ford.

MAXIMUM HOURLY VALUES

MON	MAX HR COOL	MAX HR HEAT
JAN	112,845	-19,038
FEB	114,438	-15,449
MAR	107,624	-10,868
APR	105,940	-4,739
MAY	123,642	0
JUN	133,735	0
JUL	138,024	0
AUG	122,146	0
SEP	120,032	0
OCT	124,459	-4,036
NOV	118,523	-12,907
DEC	112,524	-18,725
ANN	138,024	-19,038

TOTALS

MON	COOL LOAD	HEAT LOAD	TOTAL LOAD
JAN	80,336,345	-10,140,689	90,477,034
FEB	78,392,888	-6,739,516	85,132,404
MAR	94,761,916	-4,340,496	99,102,412
APR	101,792,160	-737,040	102,529,200
MAY	117,195,531	0	117,195,531
JUN	124,452,780	0	124,452,780
JUL	142,220,374	0	142,220,374
AUG	133,539,599	0	133,539,599
SEP	120,979,860	0	120,979,860
OCT	108,031,683	-872,619	108,904,302
NOV	87,417,210	-5,034,060	92,451,270
DEC	80,701,494	-9,707,991	90,409,485
ANN	1,269,821,840	-37,572,411	1,307,394,251

ITEM SUMMARY

MON	LIGHTING	PEOP LAT	PEOP SEN	VENT LAT	VENT SEN	ROOF	GLASS CON	GLASS SHG	WALL
JAN	71,741,471	7,502,000	7,492,452	966,580	3,037,659	0	43,328,080	37,106,628	6,314,576
FEB	64,798,748	6,776,000	6,767,376	781,200	2,379,132	0	35,384,384	36,490,552	4,634,588
MAR	71,741,471	7,502,000	7,492,452	356,500	1,964,749	0	32,039,492	41,635,666	3,589,428
APR	69,427,230	7,260,000	7,250,760	1,427,400	763,590	0	18,443,760	38,875,290	1,232,550
MAY	71,741,471	7,502,000	7,492,452	151,900	444,664	0	9,397,216	37,992,549	1,004,369
JUN	69,427,230	7,260,000	7,250,760	1,917,900	788,910	0	6,372,720	35,669,520	2,397,600
JUL	71,741,471	7,502,000	7,492,452	4,218,480	1,561,036	0	8,920,560	36,843,066	4,157,565
AUG	71,741,471	7,502,000	7,492,452	2,693,280	1,007,159	0	6,487,680	38,047,633	2,838,980
SEP	69,427,230	7,260,000	7,250,760	1,131,000	521,310	0	7,251,840	38,380,470	1,651,710
OCT	71,741,471	7,502,000	7,492,452	1,219,230	655,774	0	17,377,360	38,277,219	1,701,652
NOV	69,427,230	7,260,000	7,250,760	295,800	1,834,980	0	30,529,280	35,195,150	3,539,970
DEC	71,741,471	7,502,000	7,492,452	915,740	2,900,422	0	41,899,600	36,121,138	6,147,796
ANN	844,697,965	88,330,000	88,217,580	16,075,010	17,859,385	0	257,537,972	450,635,321	39,650,764

Table 14-3. Existing Building LOADCAL Floor Summary

DAYLIGHTING ANALYSIS

Analyzing the existing daylighting levels, within the typical office level, involved defining such factors as the typical level's geographical orientations and specific zone configurations. In the LOADCAL analysis, the typical office floor was divided into five zones, with each perimeter zone having one orientation. In this analysis, the perimeter zones are further subdivided into bays, or rooms. This is done in order to effectively determine the available daylighting.

Each bay was analyzed to determine dimensions, ceiling heights and surface reflectance (ceiling, wall and floor) values. Specific window conditions including orientation, area, transmittance value and blind usage were established. Direct sunlight could not enter the space due to the positive stop provided with the horizontal blinds. Typical adjustment angle was observed to be approximately 45 degrees.

Actual Atlanta meteorological data were used in this analysis. Solar altitude angles and azimuths were calculated for specific times and then window-to-sun orientations were established.

Illumination levels, at the windows of each bay were determined for various times of the day, under clear and cloudy sky conditions. Coefficients of utilization were established based upon the specific bay and window conditions. From these factors, daylighting levels were calculated on three work surfaces, located within each bay. Daylighting levels were calculated for the following times: 8:00 a.m., 10:00, 12 noon, 2:00 and 4:00 p.m., for March 21st, June 21st and December 21st. (Note: illumination levels for September 21st equals the levels for March 21st.)

A computer spreadsheet was developed to calculate the daylighting levels described above. The appendix contains a sample of the spreadsheet, bay number one at March 21st under clear and overcast sky conditions, used to determine the daylighting levels. The procedure is similar for the other times and for the other three bays. Following are graphs which describe the daylighting levels for the times and sky conditions noted on each graph.

The daylighting levels, which vary depending upon bay orientation, time of day and sky condition, will be the basis for evaluating any modification to the existing building. The strategy is to utilize the natural daylighting available to reduce artificial lighting levels, thereby improving the thermal performance of the structure, which will reduce the operating costs.

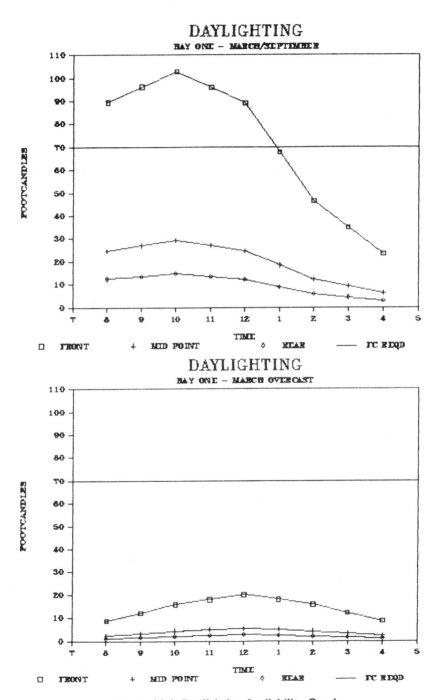

Figure 14-1. Daylighting Availability Graph

REVISED BUILDING SCHEME DESCRIPTION

As stated earlier, the goal of this investigation is to analyze the implications of using daylighting on a building's thermal load. In an effort to effectively analyze this issue, without having to consider the possible interactive effects of multiple alterations, the basic building design was not significantly altered. Changes to the overall design, such as modifications to floor height, window configuration, overhang or vertical fin depths were not considered. The addition of reflectors or sun screens was not considered either. Building usage patterns nor operational procedures could not be altered. With these constraints established, as the basis for defining alternative solutions, the only remaining element for consideration is the window glazing.

The existing glazing is a bronze-tinted, double insulating glass, "Solarban 550-60" as manufactured by PPG. It has a shading coefficient of .46, a visible light transmittance of 32 percent and a U-value (winter) of .50. It was a good glass selection, considering the types available when the facility was constructed.

Now, however, glazings with greatly improved performance characteristics are available. Low-emissivity glass and specialty coated glazings are available. One manufacturer has developed a product line called "Low - E" glass, after this new technology. The price range of these new glazings is comparable to existing insulating glass.

In selecting a new glazing type for this building, the following factors were evaluated for the types available today: the shading coefficient, the visible light transmittance and the U-value of the glazing.

The new glass should have the lowest shading coefficient factor available and the highest visible light transmittance. This will provide the least amount of solar heat-gain in relation to the amount of daylighting entering the space. By having a low shading coefficient and high light-transmittance, daylighting is provided while the heat associated with the light is filtered out.

The U-value, of course, needs to be as low as possible to cut down on the conductive heat losses/gains to/from the environment. Eliminating these adverse effects will make the space 'feel' more comfortable for cold drafts in winter and radiated heat in summer will be reduced.

The methodology for selecting a new glazing type, at the outset, was to select a glasing with a shading coefficient comparable to the existing and a visible light transmittance factor as high as possible. While this approach was sound from the standpoint that existing artificial lighting levels could be greatly reduced, the amount of natural daylighting provided was excessive and, thus unacceptable. The lighting levels generated were so great that glare and over-illumination would have made the space impossible to work in, especially at the window locations.

The opposite approach was then investigated. By selecting a glazing with a transmittance factor comparable to the existing and the lowest available shading coefficient, the results were more acceptable. While the availability of natural daylighting did not change significantly (it was actually decreased by six percent from the levels available in the existing design), the solar gain associated with the daylight was decreased dramatically.

The glazing selected is manufactured by the Interpane Group, and comes from their IPASOL solar reflective, insulating glass line. The glass is the IPASOL Gold 30/17. It is a double-insulating glazing with a gold-tinted low-emissivity coating applied to the number three surface. The visible light transmittance is 30 percent, lower than the existing glass by six percent. The shading coefficient factor is .19, which is 59 percent lower than the existing. The U-value (winter) is also much lower at .23 compared to .50 for the existing.

This one modification has the potential of providing considerable energy savings, if only by improving the thermal performance of the glazing portion of the building. However, when accounting for the artificial lighting level reductions possible by incorporating the available daylighting, an even greater thermal load reduction is probable.

```
            Existing Glass:
U-value (summer)        = .59
U-value (winter)        = .50
Shading Coefficient     = .46
Visible Light Transmission = 32%

            Glazing Selection One:
U-value (summer)        = .33
U-value (winter)        = .32
Shading Coefficient     = .42
Visible Light Transmission = 61%

*       Glazing Selection Two:
U-value (summer)        = .23
U-value (winter)        = .23
Shading Coefficient     = .19
Visible Light Transmission = 30%
*  = Glazing selected for Building Modification.
```

Table 14-4. Glass Comparison Data

REVISED DAYLIGHTING ANALYSIS

Daylighting data were developed for the revised building scheme. This procedure involved factoring the existing building daylighting values by the six percent reduction of the transmittance factor.

Once these values were determined, an average illuminance availability was calculated for each bay. This was done for the times, months and sky conditions described in the earlier daylighting analysis. Yearly illumination levels were then established for each bay. These yearly values were factored by a 'usage factor' based on daylighting availability in relation to lighting requirements, in terms of hours per day. Then, these yearly averages were compared against the lighting level requirement of 70 footcandles.

From these calculations, reductions in the existing artificial lighting level, expressed as percentages, were determined. A 15 percent reduction in the artificial lighting level is possible within bays one and two. In bays three and four, a nine percent reduction is possible.

It must be noted that, while incorporating the available daylighting will allow for reduced artificial lighting levels, they cannot be permanently reduced. A control device, or devices, will be required to increment the amount of artificial lighting reduction, based upon the contribution of daylighting at any given time. This is due to the ever changing nature of daylighting.

REVISED LOADCAL ANALYSIS

Data concerning the modified building, the new U-value, the new shading coefficient and the reduced artificial lighting levels, were entered into the LOADCAL program for analysis. All other building usage and design factors remained as described in the original analysis. The results, in terms of reduced thermal load, are impressive.

The results are described in the following chapter. The LOADCAL analysis data will be contained in the appendix. It will be presented in the same format as provided for the original building analysis.

RESULTS

By modifying the existing building with the new glazing and incorporating the available daylighting, to reduce the artificial lighting level, significant reductions in the estimated energy consumption are realized.

The much-improved U-value (winter) of .25 of the new glass yielded a reduction of 53.99 percent in the glazing conductive heat transfer. While being significant in terms of the amount of actual heat loss/gain, it will also have the effect of increasing the comfort level of the space. This is accomplished by cutting down on cold down-drafts near the windows in the winter and conductive heat gain in the summer.

The lower shading coefficient provided a 58.95 percent reduction in the solar heat gain to the space. This is a considerable reduction, especially when one realizes that the amount of natural daylighting entering the building was reduced only six

MAXIMUM HOURLY VALUES

MON	MAX HR COOL	MAX HR HEAT
JAN	61,931	-7,929
FEB	63,180	-5,719
MAR	61,823	-3,361
APR	71,363	-80
MAY	80,811	0
JUN	87,352	0
JUL	91,775	0
AUG	83,854	0
SEP	74,001	0
OCT	70,930	0
NOV	66,033	-4,578
DEC	61,964	-7,753
ANN	91,775	-7,929

TOTALS

MON	COOL LOAD	HEAT LOAD	TOTAL LOAD
JAN	67,134,003	-2,787,923	69,921,926
FEB	64,161,188	-1,583,624	65,744,812
MAR	76,496,995	-698,647	77,195,642
APR	80,936,670	-2,400	80,939,070
MAY	92,085,097	0	92,085,097
JUN	96,374,370	0	96,374,370
JUL	108,405,822	0	108,405,822
AUG	102,503,298	0	102,503,298
SEP	93,611,730	0	93,611,730
OCT	86,604,824	0	86,604,824
NOV	71,981,040	-942,900	72,923,940
DEC	67,564,717	-2,625,297	70,190,014
ANN	1,007,859,754	-8,640,791	1,016,500,545

ITEM SUMMARY

MON	LIGHTING	PEOP LAT	PEOP SEN	VENT LAT	VENT SEN	ROOF	GLASS CON	GLASS SHG	WALL
JAN	64,203,697	7,502,000	7,641,190	966,580	3,037,659	0	19,931,264	15,249,272	6,314,576
FEB	57,990,436	6,776,000	6,901,720	781,200	2,379,132	0	16,278,192	14,982,520	4,634,588
MAR	64,203,697	7,502,000	7,641,190	356,500	1,964,749	0	14,739,570	17,101,708	3,589,428
APR	62,132,610	7,260,000	7,394,700	1,427,400	763,590	0	8,488,020	15,943,380	1,232,550
MAY	64,203,697	7,502,000	7,641,190	151,900	444,664	0	4,323,384	15,583,111	1,004,369
JUN	62,132,610	7,260,000	7,394,700	1,917,900	788,910	0	2,931,480	14,612,070	2,357,600
JUL	64,203,597	7,502,000	7,641,190	4,218,480	1,551,035	0	4,102,574	15,116,576	4,157,535
AUG	64,203,697	7,502,000	7,641,190	2,693,280	1,097,159	0	2,964,246	15,558,146	2,838,380
SEP	62,132,610	7,260,000	7,394,700	1,131,900	521,310	0	3,336,240	15,795,900	1,651,710
OCT	64,203,697	7,502,000	7,641,190	1,219,230	655,774	0	7,994,156	15,728,811	1,701,652
NOV	62,132,610	7,260,000	7,394,700	235,800	1,834,980	0	14,089,860	14,461,440	3,989,970
DEC	64,203,597	7,502,000	7,641,190	915,740	2,900,422	0	13,874,560	14,831,051	6,147,796
ANN	755,946,755	88,330,000	89,968,850	16,075,010	17,859,385	0	118,473,946	185,005,385	39,660,784

Table 14-5. Revised Building LOADCAL Floor Summary Data

percent. (The visible light transmittance is 30 percent for the new glass, while the existing glass was 32 percent.)

The reduced thermal load afforded by lower artificial lighting levels, is 10.51 percent. While this may not seem significant, when compared to the reductions above, it is a tremendous amount of energy considering the overall lighting level of 70 footcandles is still provided at each work surface. And when one realizes the amount of electrical consumption saved, by the artificial lighting reduction, the actual energy savings are significant.

Another significant reduction that this modification affords is a 33.57 percent reduction in the peak-hourly cooling load of the space. While this will save considerably in the overall energy consumption, it will yield great economic savings on the amount of peak-electrical demand charges applied to the electrical consumption. It could also provide considerable construction cost savings since all mechanical equipment must be sized to handle the maximum hourly cooling load. Should the owner decide to install the new glazing as a retrofit program and reduce the artificial lighting levels within the building, additional savings could be realized by rebalancing the mechanical system. The greatest potential savings would, however, result in applying this analysis to the design of a totally new facility.

Finally, the total thermal load reduction afforded by these modifications is 22.25 percent. This is quite a significant reduction when one considers the simple modifications that were suggested.

Discussion of the Results:

The amount of thermal load reductions and economic savings provided are significant by any standard. This is especially true considering that the suggested modifications, to the existing building, are very minor compared to methodologies typically employed to reduce energy consumption levels of a structure. As noted earlier, the cost of the new glazing is comparable with other insulating glazings, and when factored for inflation, not significantly more expensive than the existing glazing in the building.

The economic justification, for making this modification to the facility, has been proven with the reduced thermal loading, lower electrical consumption rates for artificial lighting and cooling loads, reduced peak-load electrical surcharges and reduced mechanical plant size requirements.

CONCLUSIONS

In conclusion, this analysis proves that great reductions in a structure's energy consumption can be achieved with little change in its appearance or construction. It is possible to achieve the desired results of a more energy efficient building without significantly increasing the initial construction costs. The results of this analysis also indicate any increase in the initial construction costs can be recovered in the reduced operating costs of the facility.

Furthermore, this study indicates that initial choices and decisions in the design phase directly affect the overall energy performance of a structure. It is imperative, therefore, that architects and engineers analyze design choices and decisions for their possible effect upon the structure's thermal performance. It is also important that owners and developers realize that initial construction costs are not always the most important factor in a building's design. The performance characteristics and operating costs should be addressed. As energy prices begin their upward spiral, as predicted, the service of providing energy efficient structures with lower operating costs will become even more important.

RECOMMENDATIONS

The analysis procedure outlined in this chapter, while somewhat lengthy, would not be too burdensome to incorporate into the initial design phase of a structure. The author believes that it can serve as an effective analysis tool to evaluate proposed designs and/or modifications to facilities such as the one described herein. It is recommended that interested persons utilize this procedure, as another design determinant, in the overall design process.

It is also recommended that further research into this aspect of the design process be undertaken to streamline the procedure outlined within. This would be a valuable service to all parties interested in providing energy efficient structures and facilities. Finally, it is hoped that further research into technologies and methodologies be conducted that will allow even more energy efficient structures to be designed and constructed.

References

1. American Society of Heating, Refrigerating and Air-Conditioning Engineers, Inc.; *Cooling and Heating Load Calculation Manual;* New York; ASHRAE; 1979; Second Printing.

2. American Society of Heating, Refrigerating and Air-Conditioning Engineers, Inc.; *Handbook - 1977 Fundamentals;* New York; ASHRAE; 1979; Second Printing.

3. American Society of Heating, Refrigerating and Air-Conditioning Engineers, Inc.; *Handbook - 1978 Applications;* New York; ASHRAE; 1979; Second Printing.

4. American Society of Heating Refrigerating and Air-Conditioning Engineers, Inc.; *Handbook - 1979 Equipment;* New York; ASHRAE; 1979; Second Printing.

5. American Society of Heating, Refrigerating and Air-Conditioning Engineers, Inc.; *Handbook - 1977 Systems;* New York; ASHRAE; 1979; Second Printing.

6. Libbey-Owens-Ford Company; *How to Predict Interior Daylight Illumination;* Ohio; LOF; 1976.

7. Jarrell, R. Perry; "Special Problems Research: A comparison of Computer Estimation Programs"; Class Paper; 1985.

8. Aeck Associates, Architects; "Contract Documents for the Construction of Twin Office Towers, Project No. GBA-39, Georgia State Financing & Investment Commission"; Atlanta, Georgia; 1977.

9. Georgia Building Authority (GBA) "Construction, Energy Consumption and Occupancy Records for the Twin Office Towers"; Atlanta, Georgia; 1984.

10. Akridge, Professor Max; et. al.; "Class Notes for the Energy in Architecture Series"; Georgia Institute of Technology; 1984.

Chapter 15
Windows and Daylighting*

S.E. Selkowitz, D. Arasteh, D.L. Dibartolomeo,
A.J. Hunt, R.L. Johnson, H. Keller, J.J. Kim,
J.H. Klems, C.M. Lampert, K. Loffus,
K. Papamichael, M.D. Rubin,
M. Spitzglas, R. Sullivan,
P. Tewari, and G.M. Wilde

Approximately 20% of annual energy consumption in the United States is for space conditioning of residential and commercial buildings. About 25% of this amount is required to offset heat loss and gain through windows. In other words, 5% of U.S. energy consumption—the equivalent of 1.7 million barrels of oil per day—is tied to the performance of windows. Fenestration performance also directly affects peak electrical demand in buildings, sizing of the heating, ventilating, and air-conditioning (HVAC) system, and the thermal and visual comfort of building occupants.

The aim of the Windows and Daylighting Group of Lawrence Berkley Laboratory (LBL) is to develop a sound technical base for predicting the net thermal and daylighting performance of windows and skylights. The group's work will help generate guidelines for design and retrofit strategies in residential and commercial buildings and will help develop new high-performance materials and designs.

One of Lawrence Berkley Laboratory's (LBL) program strengths is its breadth and depth: at one extreme, LBL can

*This work was supported by the Assistant Secretary for Conservation and Renewable Energy, Office of Buildings and Community Systems, Building Systems Division and Office of Solar Heat Technologies, Solar Buildings Division of the U.S. Department of Energy under Contract No. DE-AC03-76SF00098.

examine energy-related aspects of windows at the atomic and molecular level in its materials science studies, and at the other extreme, LBL can perform field tests and *in-situ* experiments in large buildings. They have developed, validated, and use a unique, powerful set of computational tools and experimental facilities. Their scientists, engineers, and architects work in collaboration with researchers in industry and academia.

To be useful, the technical data developed by LBL's program must be communicated to design professionals, industry, and other public and private interest groups.

LBL research is organized into three major areas:

- Optical Materials and Advanced Concepts

- Fenestration Performance
  - Thermal analysis
  - Daylighting analysis
  - Field measurement facility
  - Building monitoring

- Building Applications and Design Tools
  - Nonresidential studies
  - Residential studies
  - Occupant impacts
  - Design tools

DAYLIGHTING STUDIES

Providing daylight to building interiors is one of fenestration's most important functions, both from an energy perspective and from an occupant's point of view. LBL conducts a wide range of activities to establish the facilities, tools and data to address these perspectives.

Daylighting Optics of Complex Glazing and Shading Systems

A quantitative understanding of the solar-optical properties of fenestration systems is essential to accurately calculate daylight illuminance levels, glare potential, solar heat gain, and thermal comfort. Existing models are adequate for determining the properties of simple, but not complex, fenestration systems.

LBL wants to develop a methodology to analyze the luminous and solar heat gain performance of any complex fenestration system. They determine daylight performance by treating fenestration systems as light sources of varying candlepower distribution that can be calculated from bidirectional solar-optical properties. This information can be used in a daylighting model to predict interior illuminance distribution. They derive solar heat gain by processing the directional-hemispherical solar-optical properties to get the total transmitted and absorbed radiation. For these purposes, they have developed two new experimental facilities, a scanning photometer radiometer and an integrating sphere.

The scanning photometer/radiometer is used to determine the bidirectional transmittance and reflectance of fenestration components and systems, i.e., the fraction of the incoming radiation from any single direction that is transmitted or reflected toward any single outgoing direction.

LBL uses the integrating sphere to determine the directional hemispherical transmittance of fenestration components and systems, i.e., the fraction of the incoming radiation from any single direction that is transmitted toward all outgoing directions. These two facilities are part of a Daylighting Laboratory that includes a roof-top station for measuring daylight availability, outdoor scale-model test facilities, and a 24-ft-diameter sky simulator for making scale model building measurements under controlled conditions.

LBL developed a new computer program. TRA (Transmittance Reflectance Absorptance), which calculates the bidirectional solar-optical properties of complex fenestration systems given the bidirectional solar-optical properties of their component layers. Additional software was partly implemented for determining the hemispherical-hemispherical and the hemispherical-directional solar-optical properties of fenestration systems for the CIE overcast and clear sky luminance distributions, given their directional-hemispherical and bidirectional solar-optical properties. These data will serve as input to the SUPERLITE and the WINDOW computer programs for determing daylight and thermal performance, respectively.

LBL will continue developing and documenting the methodology, concentrating on new detectors and more versatile computerized data collection system for the scanning radiometer and the required software for the manipulation of bidirectional solar-optical properties. Their goal is to entirely automate the procedures for data collection and manipulation to generate a data base for a large number of fenestration systems. This data base will be input to energy analysis programs such as DOE-2.

LBL also intends to intervalidate their experimental facilities and analytical routines.

The Sky Simulator and Daylight Photometric Laboratory

A 24-ft-diameter hemispherical sky simulator (Figure 15-1) was designed and built on the University of California's Berkeley campus in 1979. In operation since 1980, it can simulate uniform, overcast, and clear-sky luminance distributions. Sky luminance distributions are reproduced on the underside of the hemisphere; light levels are then measured in a scale-model building at the center of the simulator. From these measurements LBL can accurately and reproducibly predict daylighting illuminance patterns in real buildings and thereby facilitate the design of energy-efficient buildings. The facility is used for research, for educational purposes, and on a limited basis by architects working on innovative daylighting designs.

Daylighting Analysis

In previous years LBL developed several simplified daylighting design tools (such as the QUICKLITE program), that are now widely used in the architectural and lighting design communities. Last year they expanded the range of modeling capabilities and improved computational accuracy in SUPERLITE, their advanced program. A collaborative effort to develop a new lighting/daylighting model was also initiated.

LBL continues to test and evaluate SUPERLITE, concentrating on its ability to model complex shading systems. Their approach is to define the daylight transmittance properties of the window and shading system as a candlepower distribution function that varies with sky conditions and/or the indicence

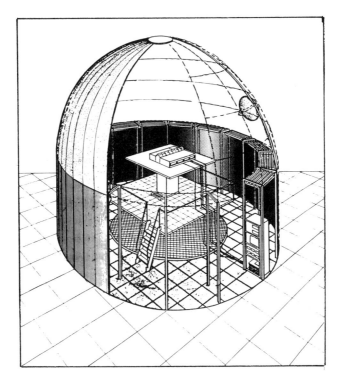

Figure 15-1. Schematic of 24-ft-diameter sky simulator with model on platform. (XBL 8412-5328)

angle of sunlight. LBL initially used theoretical distributions that could be compared to results generated by other computational techniques. This comparison showed good agreement.

LBL continues to test a version of SUPERLITE with an electric lighting modeling capability. This new capability will allow LBL to study the combined effects of daylight and electric light in a room.

LBL also worked jointly with the ABACUS group, University of Strathclyde, Scotland, to develop an improved lighting/daylighting model for their building simulation program, ESP. This model, still under development, calculates spectral data so that illuminance results can be accurately reproduced on a color computer monitor. It can also model specular surfaces. Linked to a graphics package, which displays 3-D room views, it should provide a significant advance in LBL's ability to accurately represent illuminated interior spaces.

LBL will continue testing and validation of SUPERLITE's modeling of shading devices, using candlepower data sets measured by the luminance scanner. Collaboration on development of the new ESP lighting model will continue.

Coefficient-of-Utilization
Model for Energy Simulation Models

Building energy analysis computer models must be able to predict the daylighting performance of complex design strategies commonly used by innovative architects. The models should either internally calculate the daylight illumination or be supplied with data precalculated by other illumination models or measured in scale models. The first approach, internal calculation of daylight illumination, is generally impractical for complex designs because of the significant computational cost and complexity required to obtain reasonably accurate answers. LBL is therefore developing a coefficient-of-utilization (CU) model that will be compatible with an hour-by-hour simulation model but still retain the flexibility and accuracy of more complex computational models.

LBL's approach is to derive the CU model from regression analysis of a parametric series of daylighting analyses using SUPERLITE. They previously modified SUPERLITE to generate these parametric series and also modified the output to report the indoor illuminance level due to each external light source—sun, sky, or ground.

They used a statistical computer software package to generate test regression equations for the sun, sky, and ground with a limited number of generalized variables (e.g., location of window, window dimensions). Initial results suggest that good fits can be obtained with relatively simple regression expressions. However, more extensive analysis is required to extend this to a wide range of room and fenestration conditions.

LBL will expand the effort to develop new CU models that can handle greater variation in room geometry and surface reflectance. After they generate all equations, the equations will be thoroughly tested and validated. The final model will be incorporated into future versions of DOE-2 and may also be a stand-alone illuminance model.

Daylight Availability Studies

Accurate daylight availability models are necessary for many design and energy analysis simulations. In 1978 LBL began an availability measurement project, as data were lacking for most U.S. locations. However, a widely accepted generalized model of availability in the U.S. has yet to be developed.

They previously published three papers analyzing daylight availability data for San Francisco. Analysis focused on the relationship of measured illuminance and irradiance to atmospheric parameters such as turbidity. A new functional relationship was developed to determine an illuminance turbidity for visible radiation analogous to the conventional turbidity terms used with solar radiation. LBL also developed new functional relationships for zenith luminance as a function of turbidity and found that their clear-sky luminance distribution data agree well with data from currently accepted European models. These results were published in the Technical Proceedings of the International Daylighting Conference.

In 1986 LBL's focus shifted to developing a better understanding of the nature of partly cloudy skies. Using a trailer-mounted sky luminance mapper developed by the Pacific Northwest Laboratories (PNL) they measured sky luminance profiles every few minutes throughout the day on the roof of the Space Sciences Laboratory in the Berkeley hills. Analysis of the data is being completed at Florida Solar Energy Center (FSEC).

LBL will continue to analyze their existing data base and complete a study of the luminous efficacy of daylight and sunlight. Data collection with the sky luminance scanner will be completed. They will collaborate with FSEC researchers on analysis of the availability data but will reduce the overall activity in this area at LBL.

BUILDING APPLICATIONS
AND DESIGN TOOLS

Research to develop new glazing materials and to better understand fenestration performance will provide real energy savings only if the technology is effectively applied in buildings. Using the technology requires that LBL has detailed understanding of how a wide range of fenestration systems can be optimally used in different building types and climates, and that they pass this understanding, through design tools, to building design professionals.

SIMULATION STUDIES:
NONRESIDENTIAL BUILDINGS

While most building energy simulation studies have focused on minimizing total energy consumption, other issues are equally important. Peak electrical demand affects both user costs and the utilities' required generating capacity. A complete study of the cost effectiveness of fenestration systems, particularly those incorporating daylighting strategies, must include their impact on peak electrical demand as well as on energy savings. In addition, issues such as comfort and convenience affect the user acceptance and, consequently, effectiveness of these systems. LBL's studies explore the interactions of these issues.

During the past several years the effects of a wide range of glazing properties, window sizes, lighting loads, orientations, and climates on the energy performance of a prototypical office building have been simulated with DOE-2.1B. Initially LBL examined the impact of fenestration properties, including the effects of daylighting strategies, on office building energy performance and peak electrical demand. Lighting energy savings resulting from daylighting were examined for a range of fenestration properties and lighting control systems. Annual energy consumption of an office module was found to be sensitive to variations in U-value, shading coefficient, and visible transmittance, as well as glazing area, orientation, climate, and operating strategy. Sample results from our simulation studies are shown in Figure 15-2.

LBL concludes that, in almost all instances, it is possible to find a fenestration design strategy that outperforms a solid insulating wall or roof and that daylighting is almost always an essential component of energy savings. If the installed electric lighting power density is high, the energy savings potential is large. More efficient electric lighting systems reduce daylighting benefits.

The net benefits of fenestration are highly dependent on the tradeoffs between daylighting savings and cooling loads resulting from solar gains. Visible transmittance properties, improved

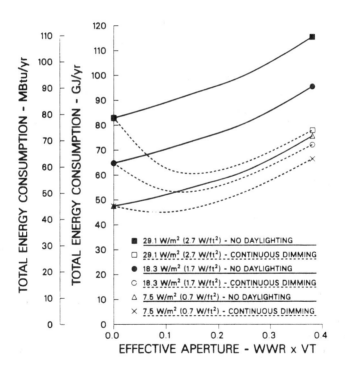

Figure 15-2. Annual energy consumption in the south zone of an office building in Madison, Wisconsin, as a function of window area. "Continuous dimming" indicates controlled reductions in electric lighting in response to daylight; "no daylighting" indicates no dimming. (XBL 876-2796)

shading design, and window management will thus assume increasing importance for maximizing energy benefits from daylight. LBL's studies have demonstrated that the common assumption that daylighting is a "cooler" source of light than electric lighting is not necessarily true. LBL has developed and is refining a methodology for comparing cooling loads imposed by daylight (or electric light) through the use of an index derived as a fraction of three parameters:

(1) The relative T_{vis} and SC of the glazing/shading system;
(2) The distribution of daylight within the space;
(3) The time-dependent absolute transmitted solar intensity.

LBL continued their general studies on the peak-shaving potential of daylighting with results that show—despite solar gains—daylighting can significantly reduce peak electrical demand during summer months. The critical tradeoffs—between electric lighting reductions from daylighting and cooling load increases from solar gain—help determine the combination of window properties that minimize building peak loads. A breakdown of this load for a sample office building at the hour of peak demand is shown in Figure 15-3 for both daylighted and nondaylighted cases. In Figure 15-4, annual peak electrical demand is shown for LBL's prototypical building in Lake Charles, LA. The benefits of daylight use are clearly illustrated.

Simulations of advanced glazing materials having active response functions, represented in the lower curve in Figure 15-5, show large potential savings.

Skylights can also provide significant energy and cost benefits. Daylighting benefits are maximized with relatively small ratios of skylight to roof areas (0.01-0.04). Because skylights provide more uniform daylight distribution, the cooling load impact of daylighting is less than with vertical fenestration. As effective aperture is increased beyond the optimum, cooling loads in most climates rise to adversely affect net annual energy performance.

A large number of DOE-2.1B runs for window and skylight studies have provided enough data for multiple regression techniques to develop analytical expressions of energy requirements

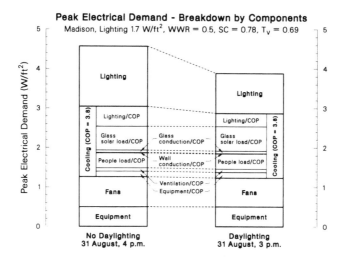

Figure 15-3. A comparison of peak electric demand, by component, for an office building in Madison, WI, with and without daylight utilization. **(XCG 855-223)**

as functions of glazing parameters; from these LBL may be able to develop a generalized expression to accommodate climate variables. The simple expressions correlate well with DOE-2 results and may become a design tool to assess energy and cost trade-offs among fenestration options. LBL has incorporated the regression equations developed in their skylight studies in a skylight design handbook sponsored by AIA, AAMA, and NFC, and this procedure is the basis for the LRI Fenestration performance Indices described in the next section.

Simulation studies to date have examined the energy impacts of many fenestration parameters for conventional designs and have begun to explore the potentials of new optical materials. New studies will examine the performance of shading devices for which adequate solar optical data do not exist. Optical properties of shading devices will be measured in LBL's laboratory. They will also continue to look at variations in window-shade energy performance and management strategies, issues of daylight luminous efficacy, advanced glazing materials, the effects of fenestration performance on HVAC, and the effects

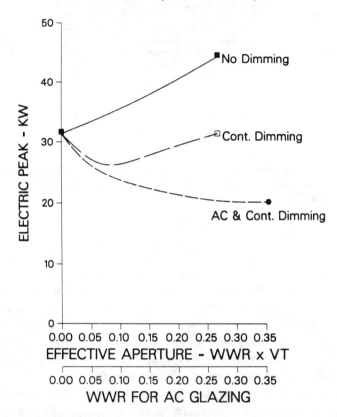

ELECTRIC PEAK FOR PERIMETER ZONES
LAKE CHARLES - 6,000 FT^2 - 2.7W/FT^2

Figure 15-4. These results from DOE-2.1B energy simulations of a 6000ft^2 office building in Lake Charles, LA, show the effect of daylighting on peak electrical demand as a function of effective aperture. Electric lighting density is 2.7 W/ft^2. (XBL 876-2802)

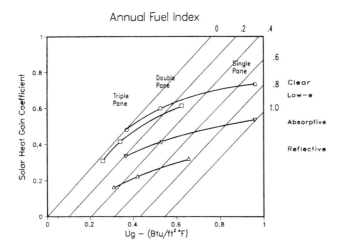

Figure 15-5. Annual fuel index as a function of glazing parameters. (XBL 876-2804)

of various HVAC options on fenestration performance. The costs of fenestration design and daylighting as influenced by peak electrical demand, annual energy use, and chiller size will be examined.

LRI FENESTRATION PERFORMANCE INDICES

Building designers, utility auditors, and others are constantly required to compare and evaluate the performance of alternative fenestration systems in buildings. To properly address the energy-related impacts of fenestration, one must be able to quantify energy performances for all systems in a systematic and reproducible way. The objective of this study for the Lighting Research Institute Center (LRI) is to develop numerical indicators of fenestration system performance on the basis of annual energy consumption, peak electrical demand, illumination performance, and thermal and visual comfort. These indicators are to be used as guides in evaluating and selecting alternative fenestration products and systems for use in various building types and climates. In Phase 1 LBL is developing the basic methodology for determining the performance indicators; Phase 2 will support the measurement and analysis required to construct a microcomputer design tool that embodies the project's results.

Using thermal and visual comfort indices in an energy design tool is a major objective of this project. A methodology was developed that defines the relationship between fenestration characteristics, direct solar radiation, and thermal and visual comfort so that annual comfort indices can be calculated. The thermal comfort index is related to mean radiant temperature within the space. Visual comfort is based on a glare index calculated by DOE-2.

Numerous DOE-2 runs have been completed for a prototype office building in Madison, WI, and Lake Charles, LA. Simplified design charts were developed from the multiple regression coefficients obtained from these runs. Figure 15-5 shows the variation of an annual fuel index as a function of solar heat gain coefficient and window U-value. Four glazing types are shown as well as the variation in number of panes of glass. Similar figures have also been generated for indices of annual electric energy and peak electric load.

Most complex fenestration products and systems cannot be readily characterized using conventional analysis techniques, so LBL is determining the solar-optical properties and daylight transmittance/distribution functions of these systems experimentally, using the integrating sphere and the luminance/radiance scanner described previously.

To complete Phase 1, LBL will develop a methodology for incorporating the proposed performance indices into an overall fenestration system figure-of-merit. A weighting function will be provided so that users can assign specific relative weights to the indices they deem important. If, for example, a designer wishes to maintain comfort without mechanical cooling, minimizing the cooling requirement might be a priority task. Thus, the cooling and thermal comfort index might be heavily weighted and an appropriate fenestration option selected accordingly.

A workshop for industry representatives will be conducted to present the methodology and performance indices developed in Phase 1. Evaluation results from the workshop will help LBL plan the development of the microcomputer design tool that is the key product of Phase 2, which they hope to initiate by the end of the year.

SIMULATION STUDIES:
RESIDENTIAL BUILDINGS

LBL's studies have focused on techniques to simplify accurately the very complicated heat transfer processes that occur between the components of a building. They have previously shown the feasibility of isolating window systems from other building components such as envelope insulation levels, infiltration, and internal heat gains in determining building energy performance.

The regression expressions developed in past years to predict residential energy use as a function of fenestration parameters were expanded to include effects from the use of night insulation, shade management, and overhangs. For a given climate, orientation, and window size, LBL developed graphic plots that allow one to quickly evaluate seasonal performance differences in alternative existing or hypothetical fenestration systems.

In addition, LBL developed a procedure to directly compare the thermal loads resulting from different building prototype configurations independent of geographic location, to ultimately help LBL develop a design tool with broad climatic applicability.

LBL used these techniques to describe the performance of low-E windows in residences in hot and cold climates. This study analyzed the energy and cost implications of conventional double- and triple-pane windows and newer designs in which substrate, type, and location of low-E coating, and gas fill, were varied. The analysis showed the potential for substantial savings but suggested that both heating and cooling energy should be examined when LBL evaluates the performance of different fenestration systems. The study also showed the importance of considering window frame effects for the low-conductance glazing units.

Most analysis procedures describe the performance of specific alternatives and one must analyze many cases to arrive at a solution that apperas to optimize energy use. LBL developed a mathematical technique to directly calculate the window size or properties that minimize energy use or cost for a given climate and orientation. It will become part of the design tools developed next year.

LBL's ultimate objective is to develop handbooks, charts, nomographs, and computer software that will help builders, developers, and suppliers assess different window strategies. The immediate objective in 1987 was to create a prototype residential fenestration design tool that would be used to stimulate building industry interest in collaborative development of a comprehensive tool.

DESIGN TOOLS
AND TECHNOLOGY TRANSFER

To influence energy consumption trends in the United States, it is critical for LBL to package and transfer their results to other researchers and professionals. The needs and motivations of this group, including designers, engineers, building owners, manufacturers, and utilities, vary widely, so LBL uses a variety of media to reach each audience. Their activities have included developing improved daylight analysis and design tools, design assistance studies, occupant response studies, workshops, conferences, assessment reports, handbooks, and meetings with industrial and design firms and utilities. Other efforts have been designed to communicate results of their work widely to other research and development groups, educational institutions in the U.S. and abroad, and professional and industrial societies.

Design Tools

The Windows and Daylighting Program and subcontractors develop and distribute daylighting design tools to industry and educational institutions. Private-sector software firms continue to introduce new design tools for daylighting but it is difficult for a potential user to evaluate and compare them. There is thus a need: (1) for a comparative matrix identifying existing tools, and (2) for photometric data and evaluation procedures to compare these tools.

Development of a general procedure for comparing design tools predictions to a photometric data base derived from sky simulator measurements and SUPERLITE simulations continued. The Daylighting Design Tool Survey was published, describing the capabilities of more than 30 tools including nomographs, protractors/tables, and micro, mini, and mainframe computer tools. The design tools tested and disseminated in 1986 are described below.

Clear Sky Protractors. Developed with Harvey Bryan at the Massachusetts Institute of Technology (MIT), the protractors consist of a series of transparent overlays that are placed over architectural plans and elevations to determine a window's daylight contribution at any point in a room.

SUPERLITE 1.0. LBL began to get professional market evaluation using the Daylighting Network, supplemented by building industry professionals in North America and abroad.

Daylighting Nomographs. The completed manual and set of nomographs to estimate potential energy and peak-load savings in commercial buildings continued to be a popular professional energy seminar package. More than 400 copies were distributed.

Skylight Design Handbook. Using established research and professional links with the American Architectural Manufacturers Association (AAMA), and their Skylight and Space Enclosure Divison, LBL developed a Skylight Design Handbook/ Energy Design Guidelines. In a cooperative project with a professional architectural firm, they converted energy design data into an easy-to-use worksheet the design professional can use to answer basic questions about the skylight design best suited for a building. AAMA will publish and distribute the Handbook to association members and the building industry at large.

Advanced Envelope Design Tool. LBL's attention has turned during the past several years to the development of the next generation of sophisticated hardware and software tools. They envision a tool that relies heavily on imaging technologies, expert-systems software, and design process studies.

LBL has developed an Advanced Envelope Design Tool concept. A schematic illustrating the key factors of this tool concept is shown in Figure 15-6. A computer data base of

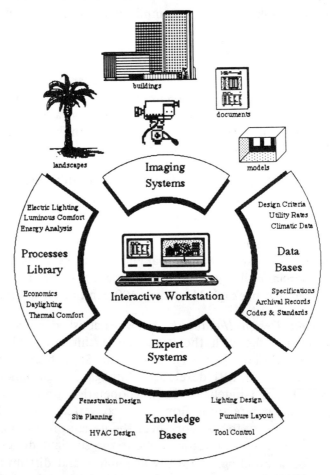

Figure 15-6. Schematic illustrating the key features of the Advanced Envelope Design Tool concept. (XBL 876-2799)

relevant hardware and software developments was established to assist in identifying and tracking market trends. Several major slide presentations were made to organizations in the building professions (AIA, ACED, ASHRAE), and a concept paper published. Planning was initiated to integrate this project with the U.S. Department of Energy's Intelligent Building Design project, to begin in 1988.

Windows and Daylighting
Building 90, Room 3111
Lawrence Berkeley Laboratory
Berkeley, CA 94720

LBL-20087: "Instrumentation for Evaluating Integrated Lighting System Performance in a Large Daylighted Office Building," M. Warren, C. Benton, R. Verderver, O. Morse, and S. Selkowitz, *Proceedings of the National Workshop on Field Data Acquisition for Building and Energy Use Monitoring,* October 16-18, 1985, Dallas, TX.

LBL-21466: "Evaluation of Integrated Lighting System Performance in a Large Daylighted Office Building," M. Warren, C. Benton, R. Verderber, O. Morse, and S. Selkowitz, *Proceedings from the 1986 ACEEE Summer Study on Energy Efficiency in Buildings: Large Building Technologies* (Vol. 3), p. 218, 1986.

LBL-21411: "Field Measurements of Light Shelf Performance in a Major Office Installation," C. Benton, B. Erwine, M. Warren, and S. Selkowitz, *11th National Passive Solar Conference Proceedings,* American Solar Energy Society, Inc. 1986.

LBL-21421: "Spectroscopic and Electrochemical Studies of Electrochromic Hydrated Nickel Oxide Films," P.Yu, G. Nazri, and C. Lampert, *Optical Materials Technology for Energy Efficiency and Solar Energy Conversion V,* Proceedings of SPIE's 1986 International Symposium on Optics and Electro-Optics, Innsbruck, Austria, April 14-18, 1986.

LBL-20347: "The Effect of Daylighting Strategies on Building Cooling Loads and Overall Energy Performance," R. Johnson, D. Arasteh, D. Connell, and S. Selkowitz, *Proceedings of the ASHRAE/DOE-ORNL Conference, Thermal Performance of the Exterior Envelopes of Buildings III,* Clearwater Beach, FL, Dec. 2-5, 1985.

LBL-20079: "Window Performance Analysis in a Single-Family Residence," R. Sullivan and S. Selkowitz, *Proceedings of the ASHRAE/DOE Conference, Thermal Performance of the Exterior Envelopes of Buildings III,* Clearwater Beach, FL, Dec. 2-5, 1985.

LBL-18234: "Transmittance Measurements in the Integrating Sphere," J. Kessel *Applied Optics*. Vol. 25, No. 16, p. 2752-2756.

LBL-20236: "Measured Net Energy Performance of Single Glazing Under Realistic Conditions," J. Klems and H. Keller, *Proceedings of the ASHRAE/DOE-ORNL Conference,* Thermal Performance of the Exterior Envelopes of Building III, Clearwater Beach, FL, Dec. 2-5, 1985.

LBL-20348: "Prospects for Highly Insulating Window Systems," D. Arasteh and S. Selkowitz, Presented at Conservation in Buildings: Northwest Perspective, Butte, MT, May 19-22, 1985.

LBL-20080: "Advanced Optical Materials for Daylighting in Office Buildings," R. Johnson, D. Connell, S. Selkowitz, and D. Arasteh, May 1986.

SECTION VII
CASE STUDIES

Chapter 16
Lighting Retrofits For A Manufacturing And Development Center

John L. Fetters, C.E.M.

INTRODUCTION

The AT&T Manufacturing and Development Center at Columbus, Ohio, was built as a Western Electric crossbar manufacturing plant in 1958. The conversion of this 2-million-square-foot facility in recent years to a modern, world-class manufacturing and software development center has provided the opportunity to convert older, inefficient lighting systems to more efficient equipment. Lighting equipment has been developed in the past few years that makes the retrofit option attractive to both energy management and lighting professionals. Replacing old or existing equipment decreases energy costs and, at the same time provides an improved workplace by improving lighting quality. The conversion of a few lighting systems will be described.

RETROFIT ECONOMICS

Several factors are used to qualify potential projects for retrofit. The age of the installation, the current operating costs of energy and maintenance, and how well the particular installation is meeting the current lighting requirements.

Lighting costs must account for the cost of the electrical energy to operate the system, the initial cost of the luminaires and lamps, the replacement costs of the lamps, and the maintenance costs. Current lighting costs at the Columbus AT&T

location divide in this way: energy costs 86%, maintenance costs 11% and replacement lamp costs 3%. This division establishes the ranking of retrofit investments with a priority for energy costs first, the maintenance costs considered next and with least attention given to the cost of the lamps.

Personal computer spreadsheets are used to calculate present costs and proposed retrofit solution savings to determine if the lighting project is economical. An example is shown in Figure 16-1. Three sections provide calculations for: 1) Energy savings, 2) Lamp savings, and 3) Labor savings. A summary at the bottom of the worksheet sums the three types of savings. The example shown is the actual cost reduction worksheet for the cooling tower retrofit described later in the chapter.

The nine cases which follow are examples of proven and successful approaches to lighting retrofits that have improved quality of life, enhanced productivity and reduced operating costs of energy and maintenance. All these cases have savings that have immediate, measurable, bottom line improvements to profitability.

INTERIOR CASES
Industrial Projects

Case 1 – High Bay Retrofit:
A 244,000-square-foot-high bay space, originally lighted by 819 twin 400 watt mercury vapor fixtures, provided an average of 23 footcandles at task surfaces. Most of the fixtures had been changed to self-ballasted lamps increasing the lighting load to 800 kW, the unit power density (UPD) to 3 w/sf, and the annual energy costs to $240,000!

The retrofit design required integrating five design criteria:

1) Increase light levels and provide good color rendition,
2) Provide a more pleasant, productive work environment,
3) Increase system efficiency and reduce operating costs,
4) Locate the retrofit system in existing pendant positions, and
5) Provide a flexible design to accommodate future changes.

COST REDUCTION WORKSHEET FOR LIGHTING PROJECTS

COOLING TOWER PROJECT

ENERGY SAVINGS ************ BEFORE LIGHTING PROJECT ************** ************* AFTER LIGHTING PROJECT **********

BEFORE: HRS/DAY= 18   DAYS/WK= 7   WKS/YR= 52
AFTER: HRS/DAY= 12   DAYS/WK= 7   WKS/YR= 52

LOCATION	#FIXTURES	WATTS/FIXTURE LAMP	BALLAST	TOTAL WATTS	HRS/YR	#FIXTURES	LAMP	BALLAST	TOTAL WATTS	HRS/YR
2 CELL TOWER SOUTH	2	100	0	200	6552	2	35	15	100	4380
2 CELL TOWER POSTS	4	200	0	800	6552	4	100	20	480	4380
3 CELL TOWER STEPS	1	150	0	150	6552	1	70	20	90	4380
3 CELL TOWER POSTS	4	200	0	800	6552	4	100	20	480	4380
4 CELL TOWER SOUTH	2	300	0	600	6552	2	50	33	166	4380
4 CELL TOWER NORTH	2	300	0	600	6552	2	400	50	900	100
4 CELL TOWER POSTS	10	300	0	3000	6552	10	100	20	1200	4380
BOIL HSE, B42, ETC	15	150	0	2250	6552	15	35	15	750	6552

TOTAL KWH BEFORE = 55037   TOTAL KWH AFTER = 16025
TOTAL KW = 8.4   TOTAL KW = 4.2

LAMP SAVINGS ************ BEFORE LIGHTING PROJECT ************** ************* AFTER LIGHTING PROJECT **********

LOCATION	#FIXTURES	LAMPS/FIX	COST/LAMP	LAMP LIFE	#LAMPS PER YR	#FIXTURES	LAMPS/FIX	COST/LAMP	LAMP LIFE	#LAMP PER YR
2 CELL TOWER SOUTH	2	1	$4.10	2450	5	2	1	$24.73	16000	0.55
2 CELL TOWER POSTS	4	1	$3.89	2450	11	4	1	$21.85	28500	0.61
3 CELL TOWER STEPS	1	1	$3.16	2450	3	1	1	$20.52	28500	0.15
3 CELL TOWER POSTS	4	1	$3.89	2450	11	4	1	$21.85	28500	0.61
4 CELL TOWER SOUTH	2	1	$5.79	2450	5	2	1	$22.10	28500	0.31
4 CELL TOWER NORTH	2	1	$5.79	2450	5	2	1	$24.78	24000	0.01
4 CELL TOWER POSTS	10	1	$5.79	2450	27	10	1	$21.85	28500	1.54
BOIL HSE, B42, ETC	15	1	$3.16	3500	28	15	1	$24.73	16000	6.14

TOTAL LAMP COSTS BEFORE = $419   TOTAL LAMP COSTS AFTER = $236

\#   3500 HR LIFE DERATED BY   30 % DUE TO TOWER VIBRATION

LABOR SAVINGS ************ BEFORE LIGHTING PROJECT ************** ************* AFTER LIGHTING PROJECT **********

LOCATION	#FIXTURES	#LAMPS PER YR	HRS/LAMP	LABOR$/HOUR	TOTAL$ LABOR	#FIXTURES	#LAMPS PER YR	HRS/LAMP	LABOR$/HOUR	TOTAL$ LABOR
2 CELL TOWER SOUTH	2	5.35	1.0	$28.38	$152	2	0.55	0.6	$28.38	$9
2 CELL TOWER POSTS	4	10.70	1.0	$28.38	$304	4	0.61	0.6	$28.38	$10
3 CELL TOWER STEPS	1	2.67	1.0	$28.38	$76	1	0.15	0.6	$28.38	$3
3 CELL TOWER POSTS	4	10.70	1.0	$28.38	$304	4	0.61	0.6	$28.38	$10
4 CELL TOWER SOUTH	2	5.35	1.0	$28.38	$152	2	0.31	0.6	$28.38	$5
4 CELL TOWER NORTH	2	5.35	1.0	$28.38	$152	2	0.01	0.6	$28.38	$0
4 CELL TOWER POSTS	10	26.74	1.0	$28.38	$759	10	1.54	0.6	$28.38	$26
BOIL HSE, B42, ETC	15	28.08	0.6	$28.38	$478	15	6.14	0.6	$28.38	$105

TOTAL LABOR BEFORE = $2,376   TOTAL LABOR AFTER = $169

SUMMARY OF LIGHTING COST REDUCTION SAVINGS

CASE 341428-071
COOLING TOWER PROJECT

	BEFORE	AFTER	SAVINGS
ENERGY COSTS	$2,752	$801	$1,951
LAMP COSTS	$419	$236	$183
LABOR COSTS	$2,376	$169	$2,207
TOTAL COSTS	$5,547	$1,206	$4,340

Figure 16-1. Cost Reduction Worksheet for Lighting Projects

A 400 watt, clear, 36,000 lumen metal halide lamp was selected to achieve the desired light output and color rendering properties. High pressure sodium sources were rejected by the owner because many of the high bay tasks are color sensitive. The commercial luminaire selected has a prismatic glass reflector with a spacing criteria of 1.7, allowing it to be mounted in existing pendant positions and still meet uniformity and light level requirements. This unique glass reflector provides 20% up light that converts the previously unused ceiling space into a luminous cavity. The resulting reflected light visually opens the formerly dark overhead space and provides a diffuse source to eliminate shadows and to redirect part of the direct component on the horizontal surfaces to an indirect component that provides the added vertical illumination needed for better visibility.

Power demand was reduced from 800 kW to 230 kW, a dramatic 70% reduction! The resulting UPD is 0.8 w/sf compared to the original 3 w/sf. First-year energy savings totalled $171,400. Now with only 498 luminaires in place of the original 819, horizontal light levels have been increased by 50%, resulting in an average of 30 footcandles on the work surfaces. The fact that the occupants have expressed the appearance of a much higher light level, suggests that the reflected light from the ceiling cavity contributes to the quality workplace beyond a simple quantity of light measure.

The use of the glass reflector has not only provided high efficiency and superior light control, but it stays cleaner and is easier to maintain. This was verified at the first group relamping. Worker acceptance has been high, resulting in lower absenteeism and higher morale. The project was paid back by savings from lower operating costs in 8 months.

The high bay lighting retrofit was awarded the Edwin F. Guth Memorial International Lighting Design Award of Merit by the Illuminating Engineering Society of North America in July, 1985.

Case 2 — Plating Room Retrofit:
AT&T manufactures several different types of telephone

apparatus, each with its own finishing specification. To ensure durability and long life, many of the metal parts are electroplated. The plating room is the action center for many plating processes. It was originally lighted with 8-foot, slimline fluorescent fixtures having porcelain-enamel reflectors. The corrosive environment of this area had taken its toll on the lighting equipment and its arrangement made it difficult to maintain. In addition, the wall and ceiling surfaces were dirty and reflected no light.

When AT&T decided to modernize this facility in 1987 it was done as a part of a plant modernization project to improve the quality of its plated parts and to provide a better place to work. A combined painting and lighting project was developed with the goal of improving product and workplace quality.

Criteria was established to: 1) Increase light levels and minimize glare from bright metal parts, 2) Use no more energy than the original lighting system, and 3) Improve access to the equipment for maintenance.

The lighting was installed in several stages, due to the limited availability and access to the room. This limitation actually helped by allowing minor equipment relocations to assure that the selected locations did not produce glare on the difficult visual tasks of handling and inspecting specular metal parts.

A mix of high-output (HO) fluorescent and metal halide sources was chosen for their efficiency and life characteristics. The fluorescent luminaire selected had a porcelain-enamel reflector with slots to allow 20% up light to take advantage of the newly painted ceiling. By using an HO fluorescent lamp instead of the slimline, 15% fewer luminaires were required to provide the design illuminance of 60 footcandles, maintained, in the north half of the room.

In the south half of the room, the three plating machines had the original slimline, fluorescent lighting embedded over the machines, as well as being located in rows parallel to the machines. These were removed and replaced with 175 watt, metal halide lamps in enclosed luminaires with glass prismatic refractors. The superior light control of this luminaire was the key to achieving glare-free lighting for viewing the metal parts. By placing the luminaires around the periphery of the machines, the machines, the tanks and their liquid contents, and the racked parts moving on conveyors can now be easily seen. Many of the visual tasks at the plating machines demand good vertical illumination and the prismatic refractor achieves this while controlling glare and surface brightness. Combined with proper placement, the result is a pleasing, quality workplace.

Two bench operations, previously employing open slimline fixtures before the retrofit were replaced with HO luminaires with acrylic, prismatic lenses. Not only was the glare controlled, but there was a noticeable difference in both productivity and quality.

A total of 13, 175-watt small optic luminaires, and 3, 250-watt, larger optic units, and 95, HO fluorescent luminaires were used for a total lighting load of 30 kW. This is slightly less than the original system which measured 32 kW. For slightly less power, the light levels have nearly tripled the values measured before retrofit; from an average of 25 footcandles (FC) on the horizontal work surfaces to an average of 70 FC. More importantly, the lighting quality has had a significant effect on the product quality. By improving the maintenance access, lighting quality can now be maintained more easily.

Case 3 – Hazardous Material Storeroom Retrofit:

Originally built as an oil storage room, this 60-foot by 56-foot storeroom was lighted with 16, 500-watt incandescent glass enclosed, porcelain-enamel, canopy reflector fixtures. The space is now used to store volatile, flammable liquids that need special handling, such as paint, solvents, and lubricating oils. With short operating hours, the energy costs were not the main concern with this system. Instead, the concern was for inadequate light levels and poor visibility for reading labels and for the safe handling of the stored liquids. A second concern was for high maintenance labor costs. Because of the volatile nature of the materials handled, an enclosed and gasketed, UL rated class 1, division 2 luminaire is required, increasing relamping time and therefore incandescent lamps.

The objectives of this lighting retrofit were to: 1) Increase light levels and improve visibility, especially on vertical surfaces, 2) Reduce maintenance and energy costs, and 3) Improve safety lighting, especially for power outage emergencies. By timing this project to coincide with a scheduled repainting, the retrofit design was able to take advantage of the improved reflectances of the walls and ceiling.

Sixteen class 1, division 2, enclosed and gasketed luminaires with prismatic glass refractors and 175-watt metal halide lamps provide a light level of 50 footcandles, maintained. The refractor directs more light downward and outward to distribute more light on the vertical surfaces and this property allowed the original mounting centers to be reused. Three of the fixtures were equipped with a quartz restrike feature for lighting while the metal halide lamps warm up. In addition, an external battery unit powers a separate sealed light at the doorway which comes on in the event of a power failure to provide a safe emergency exit. A sealed, two-lamp fluorescent luminaire was added at the storekeeper's desk.

Visibility has been improved for the safe handling of hazardous liquids and operating costs have been reduced to pay for this retrofit in less than 4 years.

COMMERCIAL PROJECTS

Case 1 – Main Cafeteria Retrofit:

The main employee cafeteria was built as a functional, institutional cafeteria, replete with silver-bowl indirect and recessed incandescent fixtures that were inefficient and expensive to maintain. The original lighting system had a UPD of 3.4 w/sf. A more modern, efficient environment that was both functional and more appealing to its employees would improve quality of life and cut operating expenses at the same time. A remodeling project was planned, with lighting to play a key role to compliment the new interior by giving each area its own kind of lighting.

The new parabolic, 3-lamp fluorescent luminaires with high efficacy, high color rendering, 3000K, tri-phosphor lamps achieve this effect in the dining room by enhancing the softer colors and textures. The dining area supplements the natural daylighting provided by the surrounding windows. Matching, smaller parabolics illuminate the lowered ceiling perimeter area.

The food service areas are highlighted with 175-watt, 3000K metal-halide recessed luminaires, set into lowered soffits, adding sparkle and attraction. Contrast is provided by lowered lighting levels in the circulation areas. At the food service counters, lighting attracts customers by highlighting and rendering the natural foods' color and enhancing the food presentation under the low-brightness, prismatic luminaires. Backlighting in the areas behind the counters matches the circulation area fluorescent lighting. In the salad bar area, 3000K fluorescent lighting mounted over the food presents an attractive contrast with the surrounding lower general illumination.

Both project budget constraints and power budget targets were achieved. The UPD after retrofit was 1.6 w/sf and energy costs are further contained by using a commercial lighting control, scheduling 14 independent zones.

Case 2 – Small Cafeteria Retrofit:

A small food service area in the north part of the building provided a vending area and tables for use by employees at break times and meal times. Like most of the manufacturing area, it was originally lighted by 8-foot, 2-lamp F96 slimline fluorescent lamps in open reflector fixtures. The lighting load for this 3200-square-foot area was 6.4 kW, for a UPD of 2 w/sf.

When AT&T decided to redecorate the area to provide a modern, attractive space and to expand food services, effective lighting was considered a key part of the new interior. The objectives of the design were to: 1) Tailor the lighting to the needs of three functional areas, 2) Reduce the UPD to 1.6 w/sf, and 3) Provide lighting to compliment the new look.

A separate vending machine area, located along the north wall was lighted to increase the attraction of the vending machines. A low ambient level of 35 FC of diffuse overhead light, provided by 2 x 2 fluorescent luminaires, equipped with energy-saving U-tube lamps, supplements the vending machine lighting. The contrast of the two lighting sources directs attention to the machines. This technique is both attractive and energy efficient at 1.2 w/sf, and is a technique used in all new vending areas at this facility.

The dining room has two sections, divided by a center counter area where food is served at meal times. Four, 175-watt metal halide luminaires equipped with prismatic lenses highlight the food with 85 FC at mealtimes. At all other times, the center area is lighted with two 2 x 2 U-tube fluorescent luminaires, with lenses matching the metal halide units, which are connected to the emergency lighting circuit for continuous operation. A timing switch controls the serving lights by turning them off automatically after they have been on a preset time.

The two dining sections are illuminated to an average of 55 FC, maintained, from high efficiency recessed downlights mounted in the new 12-foot ceiling. Compact metal halide, 150-watt, 11,250-lumen lamps with a color temperature of 4300K and a CRI of 85 provide efficiency and color rendition needed to enhance and compliment the new, cooler look. An electronic time clock turns off the dining room lighting when not in use.

The result of this retrofit is an attractive space where the energy budget of 1.6 w/sf was achieved without any sacrifice in light level or quality.

Case 3 – Toilet Rooms Retrofit:

AT&T wanted to improve the looks of its highly visible, public toilet rooms at this location, without spending a fortune operating and maintaining them. Compared to the cost of operating the rest of its 2-million-square-foot facility, the cost of operating these public spaces is small, but an important consideration here is the impression made on employees and customers alike. If people see a toilet room lighted to 100 FC, or leaking water, they may go away with the impression that we don't care how wasteful we are. This is a hard impression to correct. So, the user impressions of the facility were an important aspect of relighting the renovated spaces.

The original facilities were lighted with 2 x 4, 2-lamp fluorescent troffers, all located in a row, down the center of the room. Although illuminance measurements showed 50 FC, the central placement of the luminaires left the walls dark which resulted in a gloomy and confined feeling. The UPD was 2 w/sf and the lights burned constantly, resulting in higher-than-necessary operating costs.

The objectives for the lighting retrofit design were to: 1) Change the dark and confining space impressions to bright and spacious, 2) Reduce operating costs and restrict UPD to 1.2 w/sf, and 3) Project an image of efficiency and quality.

The lighting concept was drastically changed from a direct system to an indirect system, from uniform to non-uniform, and the design illuminance value was reduced to 30 FC. To provide the impressions of being bright and spacious, recessed fluorescent wall slot luminaires were installed along the long walls. Single, T8, 3000K, CRI 80 lamps, chosen for their efficiency and color rendering characteristics, placed end-to-end, continuously light the walls to provide an average illuminance of 30 FC. Since the lamps are recessed above the ceiling line, no lamps are visible, eliminating the objectionable troffer surface brightness and reducing glare. The wall lighting technique is especially effective above the wash basins where the reflected light from the wall mirrors supplies an average illuminance of 60 FC on the counter top.

An ultrasonic motion sensor controls the lights off when there is no occupancy. Two 4-foot lamps, one at each wall, are powered from the emergency lighting circuit to provide continuous lighting in the event of a power failure.

The objectives of the retrofit were achieved including the power budget of 1.2 w/sf, by changing the lighting concept and using modern lighting controls and equipment. The use of the walls as part of the design was the key to this effective lighting retrofit.

EXTERIOR CASES

Case 1 – Walkway Retrofit:

Lighting for the pedestrian walkways, front drive, and two small parking lots for visitors and staff was originally provided by 32, 500-watt incandescent glass globes mounted atop 20-foot poles. The system was powered by a 480-volt underground distribution wiring system and converted to 120 volts for the lamps by individual step-down transformers in each pole base. Short lamp life caused high maintenance costs, in addition to high energy costs. Before retrofit, total annual operating costs were $6,000. The symmetric, incandescent globes, mounted on wide pole spacings did not provide adequate lighting for safety and security. The retrofit design had to: 1) Provide adequate illumination for the walkway and the adjacent roadway, 2) Increase light levels to improve safety and security, and 3) Decrease operating costs. An asymmetric, prismatic glass refractor, housed in an attractive, traditional postop luminaire to compliment the existing architecture was chosen to meet the design criteria while providing a pleasing daytime appearance. Existing poles were used in their original positions, but the base transformers were removed to eliminate a source of system inefficiency. The new luminaire operates directly from the 480-volt line. A 70-watt high pressure sodium (HPS) source was selected for its high efficacy, long life and lumen output. The asymmetric distribution puts the light where it is needed and improves uniformity.

After retrofit, annual operating costs totaled $1,000, a savings of $5,000 per year. The project was paid from savings in less that 3 years.

Case 2 – Roadway Retrofit:

The roadway at the rear of the facility was lighted with 30, 150-watt mercury-vapor, cobra-head luminaires mounted on 30-foot poles on 200-foot centers. Deterioration caused higher than normal maintenance and outages became a problem. In addition, the mercury-vapor lamps seemed to burn forever, although not much light was provided on the roadway, causing concerns for safety and security.

Design criteria for the retrofit were to: 1) Increase light levels to improve safety and security, 2) Lower maintenance costs, and 3) Reduce roadway glare and provide a more modern daytime look. A high performance, prismatic glass refractor packaged in a modern roadway housing, replaced the cobra-head on existing mounting arms. A 250-watt high pressure sodium (HPS) source was chosen for higher lumen output (27,500) and lower lumen depreciation. Visibility was improved by using a refractor design that provides sufficient high-angle candlepower for uniform pavement brightness, while limiting roadway glare. The vertical burning HPS lamp distributes over 90% of its lumen output to the side so that the glass refractor can efficiently control the light away from the base of the pole without the use of a reflector.

This retrofit project was done chiefly for the benefit of improving safety and security, while replacing a deteriorating system, with some tangible benefits from reduced maintenance and energy savings.

Case 3 – Cooling Tower Retrofit:

Three cooling towers, located on the north side of the boiler house, were all originally equipped with incandescent lamps in standlights on the top of the towers. The function of the cooling tower lighting is to provide area light for evening and nighttime tower maintenance and routine operational tasks. The original systems had deteriorated and the lighting was no longer capable of providing adequate light for personnel to see well enough to do their assigned tasks at night. In addition, the operating costs of energy and maintenance were higher than necessary. Part of the high energy cost was attributed to the lack of any control to keep the tower lights off during the daytime.

The objectives of the retrofit were to: 1) Reduce operating costs, 2) Provide surfaces, 3) Improve visibility of the towers from the boiler house, and 4) Improve safety and security. A total of 25 tower and 15 boiler house building fixtures of various wattages, ranging from 100 to 300 watts, were replaced with high pressure sodium (HPS) sources in low brightness prismatic glass luminaires. HPS sources were chosen for their long life and high efficacy, resulting in reduced labor costs for lamp changes and reduced lighting energy costs. The original symmetrical standlights on the top of the towers were replaced with pendant hung prismatic luminaires with long and narrow distribution to more uniformly light the sides of the tower.

Lighting energy before the retrofit was 55,000 kWh per year. Annual labor and lamp cost savings total $2,400 and energy cost savings are $2,000 per year, paying for the project in 2 years. The economics of this retrofit are shown in detail in Figure 16-1.

SUMMARY

Nine different lighting retrofit projects have been described that have measurably improved the quality of the workplace and reduced operating expenses for the AT&T Manufacturing and Development Center in Columbus, Ohio.

These lighting retrofit projects demonstrate that good lighting practice does not have to be sacrificed for energy efficiency. Corporations like AT&T are discovering that innovative lighting solutions, that deliver both energy efficiency and high quality, result in the most effective use of both energy and human resources.

Chapter 17
Energy Efficient
Retail Lighting Design

G.C. Whalen

Energy conservation in a retail store does not have to mean a low level of monotonous fluorescent lighting which is devoid of accent lighting. By utilizing the most efficient light fixtures, ballasts, and lamps in a carefully planned design, it is possible to achieve exciting and cost effective lighting. To correct balance of good color rendition fluorescent and dramatic incandescent lighting is the key.

GOOD LIGHTING SELLS

The main objective in the retail business is to sell merchandise. To meet this objective the retailer needs to do many things well. One of the most important is the merchandise presentation which can be helped by good lighting or destroyed with poor lighting. Good lighting sells merchandise. It can provide a store an image of quality goods, a comfortable ambiance and a pleasant atmosphere.

PRINCIPLES OF GOOD LIGHTING

The uniform high level of fluorescent lighting to supermarkets or discount chains may be effective for that type store, but this sterile uninteresting lighting can be a disaster in a better department store. The use of accent lighting provides dynamic merchandise presentation, creates drama and focuses attention for the customer. The balanced use of accent lighting and general ambient lighting is the key to effective presentation. If you start with a high level of ambient light, such as a supermarket, it is practically impossible to get a spotlight to create drama on a display. A ratio of 5 to 1 accent to general lighting is ideal. For example, a store with 30 fc general lighting should provide 150 fc on display for accent lighting. The side wall merchandise should be lighted to approximately 60-70 fc to provide a comfortable field vision and overall lighting effect.

ENERGY BUDGETS

Many stores have been designed meeting the previous lighting principles and judged to have excellent lighting at 2.0 watts per square foot in sales area. The following guideline can be used to achieve the balance between the three elements; general, accent and perimeter (valance) lighting:

Type	Watts Per Square Foot
General Lighting	0.6
Accent & Display	0.8
Valance/Perimeter	0.6
Total Sales Area	2.0

Note: Highest quality stores may go to 2.4 w/sf due to higher ratio of incandescent to fluorescent.

In addition, cover and architectural lighting may add up to .04 w/sf and showcases and power/signal up to 0.5 w/sf.

ENERGY EFFICIENT DESIGN

A large midwest department store division has remodeled five stores and built four others with the following design to achieve outstanding lighting within 2.0 watts per square foot of selling area.

The stores have all white ceilings and large areas of light colored floor tile, carpet and walls to create light, bright, comfortable atmosphere. A premium cost was paid for electronic solid-state ballasts in all fluorescent fixtures with a good return on investment from energy savings. All lighting fixtures are wired by a circuit plan to control the following levels:

- emergency lights
- work lights (30% of fluorescent)
- sales area fluorescents (remaining 70%)
- display spots and valance
- outside lighting

All circuits are controlled through low voltage relays by an energy management system.

To achieve the best color rendition and optimum illumination the *fluorescents* were specified as 40 watt and 30 watt T-12 tri-phosphor 3000 degree kelvin lamps for ceiling and valance lighting. The show cases used a 3100 degree kelvin T-8 lamp also with high color rendition.

The *spotlights and flood* lamps were specified 90 watt tungsten-halogen par 38 with color temperature very near the 3000 degree kelvin fluorescents. The selection of light sources for the majority of the store was kept to a minimum in order to simplify maintainability.

The fluorescent *general lighting* consists of 2 x 4 - three lamp deep cell parabolic silver louver with 18 cells and one three-lamp electronic ballast. Fixtures are spaced on an 8 x 8 Ashlar (diamond) pattern with 128 square foot per fixture. General lighting was 35-45 fc initial (30-40 maintained). A few stores used a simular 2 x 2 fixture on 8 x 10 pattern. This fluorescent system was used in all departments except furniture (track lighting) and cosmetics/jewelry.

The *cosmetics/jewelry core* used a fixed reflectorized downlight with 90-watt tungsten-halogen floods over showcases. Adjustable accent fixtures are also used for displays on cases and back islands.

The *accent lighting* used pull-down incandescent fixtures easily adjusted from floor. Track lighting fixtures with matching housings were used at some focal wall areas to increase visual impact.

The *valances* are high with no eggcrate to increase light or create future maintenance problems. These valances use 4-foot and 3-foot strip fixtures with double lamps and electronic solid-state ballasts.

SUMMARY

Specifying and designing lighting systems for department stores over the last 12 years, since the start of the rapid energy cost escalations, has placed strong demands on manufacturers to develop fixtures, lamps and ballasts to achieve good lighting within energy budgets. They are meeting the task.

Chapter 18
Optical Reflectors for Fluorescent Strip Lighting Fixtures for Retail Stores

C.P. Quinn, F.M. Smith

Fluorescent light fixtures are commonplace throughout our environment. Fluorescent lighting is the predominant source of light in office buildings, manufacturing plants, schools, warehouses, retail facilities, hospitals, and any building requiring efficient illumination.

The advantages of fluorescent lighting, over incandescent and other sources, are well documented: efficiency, low cost, low maintenance. Even with these advantages, fluorescent lighting still represents a significant percentage of a facility's electrical energy usage. In high-rise office buildings, the lighting system commonly accounts for 35 to 45 percent of the total electrical energy consumption.

Energy-conscious building and facility managers have been aware of this fact, and have therefore installed energy-saver fluorescent lamps and ballasts in many cases. Reducing the number of lamps of fixtures has been attempted in other cases, but the visual and aesthetic effects have generally been unacceptable. What else can a building manager implement? The optical reflector for fluorescent lighting fixtures may be an answer!

OPTICAL REFLECTORS – AN OVERVIEW

Based on proven performance and results, optical reflectors have become the most exciting energy conservation products of the 'eighties. The basic concept is surprisingly simple and effective. Although reflective materials have been used with fluorescent lighting for indirect and aesthetic lighting purposes, only recently have reflective materials been used for the purpose of energy conservation. In the early 1980s, custom-engineered, mirror-quality reflectors for installation in a fluorescent light fixture were developed. Photometric testing by independent test laboratories demonstrated a remarkable 30 to 70 percent improvement in luminaire efficiency and effectiveness.

How can an optical reflector achieve such dramatic results? The answer involves material and design. First, highly reflective specular silver coating is applied to light guage aluminum sheets. The silver coating provides up to 95 percent reflectivity of incident light. Second, the mirror-sheets are formed into multi-surfaced reflectors. As a result, the reflectors redirect light that is normally scattered, reflecting that light into the room. Since the performance of the reflector is so dramatic, 50 percent of the lamps and ballasts within the original fixture can be removed, typically without a significant reduction in light levels. In addition, the reflector "images" the remaining lamps, that is, creates the illusion that the fixture contains more lamps than it does, so that the fixture does not appear to be de-lamped. The end results:

- 40 to 50 percent reduction in lighting electrical energy usage.

- 40 to 50 percent reduction in lighting maintenance costs.

- Air conditioning load reduction.

- Enhanced lighting – less glare.

- High percentage of original light levels maintained.

- Impressive return on investment (ROI) – often within 12 to 24 months.

In addition, optical reflectors for fluorescent light fixtures are UL listed, fully tested for accelerated life, and demonstrate outstanding photometrics. The low maintenance reflectors are non-mechanical, non-electrical, and easy to install.

The financial impact of a reflector system investment is equally impressive. Using conservative calculations for a facility operation that is fitted with 2-lamp, 8-foot strip luminaires, an ROI of under 24 months is easily obtained based on energy savings alone.

For example, a 2-lamp, 8-foot strip luminaire in a retail store with a total wattage of 175 watts, operating for an average of 20 hours a day, 365 days per year, at a kWh rate of $.075, will consume $95.80 of electricity annually. With the reflector installation, the luminaire wattage is reduced by half, to a total of 87.5 watts. The electrical energy savings per fixture equals $47.90 per year. The reflector installation cost of approximately $50 to $60 per fixture is recovered in under 15 months. Higher kWh rates or operating rates provide an even better return on investment.

Reflectors have been installed by one manufacturer in over 500 major facilities in the United States during the past 6 years. Many companies have taken advantage of the reflector program and have experienced outstanding results. The companies include Pacific Bell, Amdahl, Signetics, Kaiser, Ford, Hewlett-Packard, U.S. Post Office, Bank of America, the State of California, and many others.

Now that the basic concepts of the optical reflectors have been presented, let's move on to the specific application of the reflector concept to retail lighting fixtures.

RETAIL REFLECTOR APPLICATIONS – STRIP FIXTURES

Optical reflectors can be of tremendous benefit to retail stores. Every dollar saved in operating costs can equate to as much as $30 to $100 in additional sales, depending on the margin of the store.

The application of energy-saving reflectors to the standard retail strip lighting fixture, in which there is not a lens, is much more complex than the application to a standard 2' x 4' lensed troffer. Additional attention must be paid to the strip fixture for two primary reasons:

— A lens on a standard 2' x 4' recessed troffer is very forgiving—the standard lens can easily hide some of the novice manufacturer's mistakes. With an open fixture, as in retail, there is no lens to hide mistakes.

— Appearance; because the fixture typically is surface mounted and the reflector is visible to the naked eye, an attractive appearance is critical to the success of a reflector installation.

The following discussion deals with how one manufacturer, Maximum Technology, addresses these concerns, as they relate to the standard 8-foot, 2-lamp surface-mounted retail strip lighting fixture. We recognize that there are many other fixture styles in retail use; however, we have chosen to limit our discussion to this particular fixture because it is in such widespread use, and can be such a good application for optical reflectors (if designed properly). A strip fixture and reflector are shown as Figure 18-1.

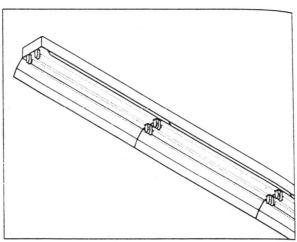

Figure 18-1. Retail Strip Fixture with Reflector.

Fixture Placement

The orientation of the lighting fixtures with respect to the aisles is a major factor to be considered in the design of reflectors. There are three main ways of orienting lighting fixtures with respect to aisles.

Parallel to Aisles — First, if the fixtures are aligned parallel to the aisles and directly over the center of the aisles, it should be clear that the type of reflector design required will be quite different than if the fixtures are aligned perpendicular to the aisles. If the fixtures are directly over and parallel to the aisles, the reflectors can be designed to direct the light in a fairly narrow distribution pattern. This will allow the level of illumination on the very tops of the gondolas to be reduced, because product isn't sold there. This will result in the concentration of light on the product area and the floor area between the gondolas.

Perpendicular to Aisles — The second scenario, in which the fixtures are aligned perpendicular to the aisles, requires that we design reflectors to provide as even a level of illumination as possible. Unfortunately, this means illuminating the tops of the gondolas, as well as the rest of the store, therefore requiring slightly more lighting power for a given illumination level. As most of you know, this arrangement appears to be the current trend in retail lighting.

Random Alignment — The third scenario is where the gondolas have been moved around after the lighting fixtures have been installed, or when the fixtures have been placed in the ceiling without regard to gondola location. This is a major concern when the lighting fixtures are running parallel to, but offset from the centers of, the aisles. In these cases, the lighting fixtures might end up directly over the center of the aisles, directly over the gondolas, or somewhere in between the two. Cearrly this presents design problems. If a reflector is designed for the wrong lighting distribution pattern, a poor lighting system will be the result. Proper engineering can compensate for these placement problems.

It is extremely important to perform a thorough lighting audit in a retail facility. This lighting audit involves superimposing a reflected ceiling plan on a gondola layout plan. Additionally, the various styles of lighting fixtures in a facility must be determined both as to quantity and location. This information makes it possible to design a reflector system specifically for a given store, to provide the best lighting for that store.

In some stores, gondolas might be moved around with time. If the placement of the gondolas in relation to lighting fixtures is uncertain, then reflectors will be designed to provide as even a distribution as possible. However, if it is certain that the gondolas will not be moved around, the reflectors can be engineered to make maximum use of the available light by directing that light from the lamps to very specific selling areas. This will provide a higher illuminating level on those specific areas, and may actually result in slightly better illumination overall.

Wall Darkening

As lighting fixtures are located near walls in retail stores, special attention must be paid to the way in which a reflector system may cause shadows on the walls. If a row of lighting fixtures parallel to a wall is fitted with reflectors that direct light down toward the selling floor, the reflectors will direct the light *away* from the upper sections of the walls, causing them to darken. In many cases this darkening will be unacceptable, because it may lend a cave-like appearance to the store.

To solve this problem, an asymmetrical reflector design should be used. This asymmetrical reflector design will be similar to a half reflector, directing light from one half of the lamp downwards, while the other half of the lamp's light will be directed towards the wall, lighting up the upper parts of the walls and any murals or artwork on the walls. Illumination of these murals and artwork is very important because of their role in attracting the customer.

Lighting the Store Entrance

If the lighting fixtures are perpendicular to the aisles and gondolas, it is likely that they will be perpendicular to the entrance of the store. An advantage to this is that when a retail customer walks or drives towards the store, the customer will see rows of fluorescent lights shining directly into his/her eyes.

A problem may arise when a reflector system is installed with a symmetrical distribution pattern to light the selling area of the store. The lamp may actually be concealed by the reflector when the fixture is viewed from afar (e.g., outside the store). This concealment keeps the lamp from shining directly into the customer's eyes, and may give the store the appearance of being closed. A specific reflector shape may be needed to expose the lamps, giving the store the appearance of being open.

One method that has been used to achieve this is to fit typically two or three of the first four or five rows of lighting fixtures from the front of the store with asymmetrical reflectors, with the open part of the reflector directed toward the store front.

Maximum Technology does not generally use this asymmetrical design on the first, second, or sometimes even the third row of fixtures. This is because these rows are typically concealed from the viewers' eyes, since only a cretain portion of the storefront is glass. In a case where the entire storefront were glass, the asymmetrical designs would start with the first row.

Some more upscale retailers are intentionally choosing the "concealed lamp" appearance. These retailers realize that they are attracting a different clientele by not having the "exposed lamp" look to their stores.

On the other hand, some retailers wish to give their stores a less expensive look, which should of course translate to a less expensive product in the customer's mind. These retailers should choose to have their customers actually see the lamps. This can be achieved by designing a very wide, shallow reflector for the installation.

One other trend becoming evident in retail stores is the tendency to have less glass area on store fronts. This makes this "concealed" versus "exposed" lamp problem less difficult to resolve.

Store Interior Lighting

Occasionally, customers have requested that we provide them

with reflectors that allow them to see the lamps of the lighting fixtures in virtually all the lighting fixtures possible. We do this by designing the reflector so that it does not come down below the center point of the lamp towards the selling floor.

There are some advantages and disadvantages to this design. The primary advantage, of course, is for retail stores wishing to create an inexpensive appearance. This type of design, which keeps the lamp fairly exposed, will give that bright, high glare appearance. Most retail buyers will interpret this to mean that the store sells less expensive products. Another advantage is that this type of design provides an extremely even distribution of light throughout the store.

One disadvantage of this type of design is that the overall lighting level is decreased somewhat below what it would be if the reflectors were engineered to provide maximum lighting without significant regard to the appearance of the reflector and lighting fixture.

Factory-Fitted Appearance
There are three reflector design characteristics necessary for a retrofit reflector to have a factory-fitted appearance:

– Full-fixture length
– White backing
– Wingtips

Full-fixture Length – The reflectors must be designed so that they run from one end of the lighting fixture to the other. Since the lighting fixture is longer than the lamp, this means that the reflector must also be longer than the lamps. This will require the reflectors to be notched at each end around the lamp holders. These reflectors will be manufactured in either two 4-foot sections or one 8-foot section, depending on the type of ballast covering: if the fixtures utilize two 4-foot ballast covers, the reflectors will most likely be manufactured in (nominal) 4-foot sections. This will allow the removal of the ballast cover and reflector in a single piece when access to the ballast is required. Otherwise, to access the ballast, the reflector would have to be removed separately from the ballast cover(s).

If the reflectors are designed to be removed with the ballast covers, they will never have to be removed separately to access the ballast. This will require that the method of attachment of the ballast cover to the lighting fixture be a tab or turnbuckle or other similar device. The use of such a device will probably require that the reflector be punched, in order that the access be unhindered. If the ballast cover is 8 feet long, then typically we will engineer the reflector in one 8-foot section.

White Backing — Maximum Technology silver reflectors are manufactured with a base aluminum which has a white backing. This white back gives the reflectors the appearance of being "original equipment." This is in contrast to the standard green-backed silver reflector and the standard green- or gray-backed polished aluminum reflector.

Wingtips — "Wingtipping" refers to the procedure of bending the outside ½″ to ¾″ of each side of the reflector upward towards the ceiling. These wingtips give the reflector a more factory-fitted appearance. Please note that these wingtips can be lint and dust traps in dirty environments. For this reason, we usually only use the wingtips in environments which are quite clean. In the dirty environments, trapped lint can actually be a fire hazard.

Color Rendition

Depending on the retail store in question, the color rendition of the products may or may not be very important. If the color rendition of the product is of extreme importance, then we typically recommend that Specular Silver Optical Reflectors (SSORs - trademark) be used. This is because the SSORs (tm) have significantly better reflection of the visible light spectrum and provide much better color rendition of the product, than do aluminum reflectors. If color rendition is not of great importance, then it might be recommended that a polished aluminum reflector be used. The polished aluminum reflector reflects about 10 percent less light than the silver reflector. It is therefore important to make sure not only that color rendition is not critical, but also that the resulting lighting level will remain adequate.

Figure 18-2 shows the reflectivity of both silver and aluminum reflectors. As you can see, the silver reflector is very good for reflecting light throughout the visible spectrum, dropping off noticeably in the ultraviolet range, while the aluminum reflector has reasonably even reflectance throughout a significant portion of the blue and violet range, and then drops off only slightly in the ultraviolet range.

If a retailer is using a high color quality lamp, then we recommend that full advantage be taken of that more expensive lamp, by utilizing a reflector that reflects the highest quality color back down towards the selling area, where it will provide the maximum positive effect on the products being sold.

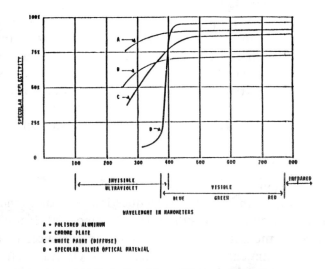

A = POLISHED ALUMINUM
B = CHROME PLATE
C = WHITE PAINT (DIFFUSE)
D = SPECULAR SILVER OPTICAL MATERIAL

Figure 18-2. Reflectivity - Silver and Aluminum.

A reflector system is not a straightforward application for retail strip lighting fixtures. It is very important to understand both the selling environment as well as the coloring and performance characteristics of a given lighting fixture/reflector combination in a specific retail store to be able to engineer a reflector system properly for a given end user.

BENEFITS TO
THE RETAIL STORE OWNER/MANAGER

There are many reasons for reducing lighting costs in a retail store. One of the most significant reasons is *profit*. If one can reduce energy costs by one dollar, and there is a one percent margin on gross sales for the store, then that reduction of one dollar in operating costs is equivalent to having increased retail sales by $100.

Suppose that a store uses $30,000 per year in electrical energy to light the store, plus an additional $5,000 per year to air condition that store, for a total of $35,000. If the cost of that lighting system and air conditioning system can be reduced by 50 percent, a savings of $17,500 will be realized. At a one percent margin, that is equivalent to having increased the gross sales of that store by $17,500 x 100, or $1.75 million in increased sales in one year. It is very easy to reduce the operating costs by a significant amount of money by using reflectors. This may be one of the easiest ways of increasing gross profits available to the retailer.

Additionally, maintenance cost savings can be realized. This is because in a typical reflector job, one lamp is removed from a two-lamp fixture. This means that only one lamp needs to be replaced in the future when the store is re-lamped. This translates to a savings of roughly $2.50 to $3.00 per fixture on lamps every 2 to 3 years, plus associated ballast and labor savings. All of this adds up to significant operating cost savings for the retailer who implements a reflector project.

SUMMARY

Fluorescent lighting for retail store is expensive. An 8-foot, 2-lamp fixture can easily cost more than $100 per year to operate. Installation of an optical reflector system can be a very effective way to decrease these operating costs, and increase store profitability.

The implementation of a reflector system is not at all easy. There are many details which must be attended to, in store layout and atmosphere and reflector design, to ensure a successful reflector project.

Chapter 19
Lighting in Schools

Anil Ahuja

BACKGROUND

Electrical bill for the Los Angeles Unified School District runs at about $24 million annually. This represents 70% of the total energy and water bill. Lighting and air conditioning costs are evenly split. In the lighting area alone, there is a huge potential to save money from energy efficient lighting and at the same time improve the quality of light in most cases.

Even though efficient lighting plays a major role in improving building operating efficiency, not much has been done in schools as commercial building standards do not directly apply to school construction, a cohesive policy on building standards is absent and efficient systems require high initial investment.

Los Angeles Unified School District has over 12,000 buildings which require adequate lighting. A recent survey revealed that 18,700 classrooms, 7,500 kitchens and offices, 16,500 auditoriums and cafeterias and 300 gymnasiums require replacement of lights with energy efficient lights. It will cost $84.5 million to replace lights in 18,700 classrooms alone.

Goal

Goal of school lighting is the creation of an effective total visual environment for learning, dependent on a three dimensional pattern of brightness and colors. It takes into account emotional and aesthetic values. The District emphasis is not only in cost-effective efficient lighting but it also addresses the need for effective lighting. Being efficient shows how well lighting equipment converts the electrical power input to light output which alludes to the relationship between output and input. Being effective requires the use of the right kind of equipment for a specific task which would promote a conducive work environment.

Criteria for Lighting Changes
As replacement of all the lights will require huge resources, a criteria for replacment of lighting has been developed as under:

1. Classrooms and libraries
2. Offices
3. Gyms
4. Auditoriums and cafeterias
5. Other, such as storerooms

Amongst the classrooms, following priority ranking has been developed:
- Classrooms with incandescent lighting
- Classrooms with hazardous/obsolete fluorescent fixtures
- Sites with high student density

Lighting Types
There are four types of lighting in wide use at Los Angeles Unified School District:

1. Incandescent lighting has the lowest efficacy (9-21 lumens/watt) and shortest lifetime (750-2000 hours). It is widely used due to low initial cost and superior color rendition.

2. Fluorescent lamps have good efficacy (65-85 lumens/watt) and a long lifetime (10,000-20,000 hours). Their color rendition is acceptable to most people. They are widely used in classrooms and office settings.

3. Metal halide lamps have good efficacy (65-80 lumens/watt) and lifetime (7,500-20,000 hours). Good color rendition makes this lamp an excellent candidate for gym and stadium lighting.

4. High pressure sodium lamps have highest efficacy (80-105 lumens/watt) and a very long lifetime (12,000-24,000 hours). Poor color rendition limits their utility to parking lots and outdoor security lighting.

Lighting Design Criteria

New Construction: The school district has established a policy to provide adequate and quality lighting in schools. This policy contributes towards aesthetics, pleasant surroundings and reduction in operating costs. It promotes an effective lighting design which must balance many different and sometimes conflicting criteria. It ensures that the design engineer does not ignore the criteria and provides efficient and effective lighting. Serious attempts are made to include the concerns of both energy management and lighting professionals.

Motion sensors have been included in all new construction design. These controls allow the maximum wattage per square foot to increase and thus provide greater flexibility in design within the interior lighting power allowance under the Title 24 system performance method.

Progress is being made to coordinate the design of HVAC, electrical and lighting experts to ensure that whole building energy allowance is met.

We are considering the use of luminaires in lighting design. By having the photometric data of several lighting equipment manufacturers stored in computer memory, analysis of the proposed design for quality lighting becomes a simple matter. A perspective graphic package enhances this capability.

Retrofit: The school district requires that lighting level of 70 foot candles from fluorescent lighting be maintained except the following major areas:

Locations	Foot Candles	Light Source
Drafting, sewing conference, and sight-saving rooms and shops	100	Fluorescent
Corridors	20	Fluorescent
Auditoriums	30-70	Incandescent
Gymnasiums	30-70	Metal Halide
Athletic Field (Stadium)	45	Metal Halide
Parking Lots	2+	High Pressure Sodium
Building Exterior (Security)	35-100 watts	High Pressure Sodium

Few Pointers on Fluorescent Lighting

— Natural lighting: Use as much natural lighting without glare to provide high visual comfort probability (VCP). Glare is minimized by adjusting the sun angle.

— Refractors: Light at a particular angle of refraction provides exceptional brightness control. The prismatic lens create uniformity of light cutting down the glare.

— Prismatic sheilding: Provides optimum efficiency.

— Diffusers: Help to spread light and provide uniform and soft illumination by eliminating shadows.

— Surface mounted/flush mounted fixtures: Both types are capable of providing superior photometric performance. However, surface mounted offers ease of installation and maintenance and flush mounted gives out less glare and a more refined appearance and may require higher number of fixtures.

Terms to Learn

— Maintenance factor (MF) = Luminaire Dirt Depreciation (LDD) factor x Lamp Lumen Depreciation (LLD) factor.

— Coefficient of Utilization (CU): Directly proportional to the amount of reflectance.

— Foot Candles = $\dfrac{\text{Number of Lamps x Lumens per lamp}}{\text{Area}} \times \text{MF} \times \text{CU}$

CASE STUDIES ON RETROFIT

Case Study 1

Stadium lighting at Birmingham High School

Criteria for Selection: High energy usage

High potential for energy savings

Public relation value (Mayor's Stadium)

Old System

160 incandescent lamps of 1500 watts each = 240 kW

Light levels measured on field = 12 Footcandles

> 110-foot-high poles made maintenance difficult
>
> Resulting in high number of burned-out lamps

Lamp life expectancy = 750 hours

New System

60 super metal halide lamps of 1000 watts each = 60 kW

New light levels = 35 Footcandles

> This light level calculated at mean lumen output of lamp, rather than initial lumen output.

Lamp life expectancy = 12,000 hours

Savings Per Year

240 kW − 60 kW = 180 kW x 1664 hours	= 299,520 kWh
Energy Cost @ 7 cents per kWh	= $20,966 year
Replacement Cost	= $5,984 (labor on boom truck)
Total Savings	= $26,950
Cost of the Project	= $33,000
Payback	= 1.2 years

Additional benefits: Safety, replacement of two PCB transformers with one, yielding savings of $50,000 and improved lighting level.

Source of Funds

- Utility budget
- Maintenance
- Student body funds
- Operations
- School budget (IMA)

Case Study 2

> Gymnasium lighting

Old System

30 incandescents at 1000 watts each = 30 kW

New System

30 high pressure sodiums at
400 watts each   = 12 kW

Savings/Year   = 18 kW x 3600 hrs. = 64,800 kWh

                   = 64,800 kWh x 7 cents/kWh = $4,536

Cost   = $7,000

Payback   = 1.5 years

Poor color rendition from high pressure sodiums has forced the District to consider metal halides for future projects.

Special Applications

Visually Handicapped Schools: Many low-vision children have special and unusual sensitivity to illumination levels. Most commonly strong lighting is preferred but in some instances dim illumination is required for best performance. Thus there is a need to ensure flexibility to cater for the wide variety of lighting needs.

Many commonly used fluorescent tubes have a disproportionate amount of their luminous energy in the blue region of the spectrum and this needs to be avoided. Many low-vision children will have some haziness of the normally clear media of their eyes; because blue light is scattered more than longer wavelengths, it is disadvantageous to use blue rich lighting when there is haziness of the ocular media. Incandescent or red-rich fluorescent is more desirable in such cases.

Daylighting: This is the best source of light if there are no glare problems. Blinds or other light shieding can remove glare whenever necessary. Further control of light can be achieved with automated dimmer controls. Application of these controls to fluorescents is in its infant stage but could bring about promising results to balance the light in a classroom and at the same time reduce electrical lighting costs.

Chapter 20
New Lighting Options
For the State Buildings

Jorge B. Wong-Kcomt,
Wayne C. Turner, Shuibo Hong,
William R. King

INTRODUCTION

A large energy management project is underway where Oklahoma State University and the Office of Public Affairs for the State of Oklahoma are working together to reduce energy costs for state buildings under the control of Office of Public Affairs. One of the biggest cost components is lighting and the vast majority of lighting is standard fluorescent tubes. Much technological development has occurred recently in fluorescent tubes so there are many retrofit options.

The Energy Division of the Office of Public Affairs wanted to find out which options can be cost effective and, at the same time, be accepted by the building occupants. This chapter presents 1) the conclusions from an economic analysis of the most relevant fluorescent lighting alternatives for the State Buildings and 2) the results from an experimental test carried out in an occupied building. The economic evaluation has been performed by using a life cycle approach and the computation for the costs and savings of each option has been done on the measures of energy usage and readings of light levels on the tested options are included.

Significant energy cost reduction will be realized when the recommended lighting options are fully implemented in all buildings of the State Capitol Complex, Oklahoma City. These recommendations can be extended to all state buildings of the State of Oklahoma.

The Office of Public Affairs (OPA) of the State of Oklahoma in conjunction with the School of Industrial Engineering of Oklahoma State University (OSU) is carrying out a large energy management project for the state office buildings. An important part of this project constitutes both the technical and economical evaluation of more efficient fluorescent lighting systems.

A variety of new lighting technologies have entered the market. In most existing buildings and facilities, when confronted with specifying a fluorescent bulb and/or accessories for energy efficiency, building and plant managers are usually confused with such diversity.

Considering that many of the actual lighting systems were not designed for energy conservation, the new options for fluorescent lighting may lead the state buildings to opportunities for energy and dollar savings. A selected set of these opportunities is presented in this chapter.

To evaluate the actual energy savings attainable from some of these options, and to gather the opinions of building occupants about the new options, the OSU research team conducted an experiment in one of the State buildings. Results of this experiment are presented as well.

Because of configurations of space or use, some options could be more advisable, even though they are less efficient. For instance, historical buildings and art exhibits usually require lamps with a minimum emission of ultraviolet rays to prevent artwork deterioration.

AVAILABLE TECHNOLOGIES

Among the products offered in the market as "fluorescent lighting energy savers," for a four-40-watt-lamp fixture (2 ft x 4 ft), we have selected four different categories:

1) Energy efficient fluorescent lamps, EEFL
2) Electronic fluorescent ballast, EFB
3) Power reducing devices, PRD
4) Improved fluorescent reflectors, IFR

Energy efficient fluorescent lamps consume less energy than standard (F40) lamps while giving nearly the same light levels. They cost more initially, but the incremental cost will be recovered through energy savings. Essentially, there are two kinds of EEFL, 1) those that do not alter the color rendition and visual definition, and 2) those that have a higher lumen output and an improved color rendition. Within the first kind of EEFL we consider energy efficient (EE), energy efficient "plus" (EE+) and Octron (T8) lamps. Some examples of new lamps with improved color rendering illumination and higher lumen rating are the Aurora IV from VL Service Lighting Corporation, and the Advantage X from North America Philips Lighting Corporation.

Ballasts designed with solid-state electronics have recently become commercially available. Their high frequency operation (near 25,000 Hz) allows fluorescent lamps to operate more efficiently and still provide a light level equal to that of traditional electromagnetic ballasts. Some energy efficient lamps, as the Octron, require special magnetic ballasts, but the manufacturer recommends the use of electronic ballasts to maximize energy savings.

Power reducing devices basically "regulate" the flow of current going into the bulb once the lamp is lit. These devices can also reduce the lumen depreciation rate, allowing a longer bulb life. One example of these devices is the Edison 21 fluorescent monitor. The Edison 21 comes in two variations. One saves 30% and the other 50% of energy consumption.[1] A variation of PRDs is a "power reducing fluorescent bulb," marketed under the commercial name of "Thrift/Mate" by Sylvania. Two Thrift/Mate lamps are required in four-lamp fixtures, one on each ballast circuit. This lamp accomplishes the same purpose as a PRD. One type of Thrift/Mate reduces power and energy by 33% and another type by 50%. The manufacturer claims than lumens-per-watt efficiency remains the same after replacement. Hence, illumination levels are reduced proportionally to the power/energy reduction.[2] Therefore, this option may be used in rooms with excessive illumination. This is the case in many Oklahoma State buildings. In general, PRDs may be considered as an alternative to EEFL.

Fluorescent reflectors can be improved by using surfaces or films with better reflectivity. In addition, the position of the reflective surfaces with respect to the bulb may affect the performance of a given fixture. Improved lighting levels can be achieved with a good combination of high reflective surfaces and well designed reflector geometry. A vendor claims a "completely lit" impression for a four-bulb fixture, when two lamps are removed, the two existing lamps are relocated and an "optimum bend reflector" is installed in the fixture.[3] In this chapter we consider only reflectors for four-lamp fixtures.

ECONOMIC ANALYSIS

From the four categories mentioned above, eight alternatives have been considered as "challengers" to the "do-nothing" option, e.g., keep using a 4-tube fixture with 4 standard 40-watt lamps and standard ballasts. The alternatives are listed as follows:

1. Use 4 energy efficient fluorescent lamps (EE) and keep standard ballasts.
2. Use 4 EE lamps and electronic ballasts.
3. Use 4 energy efficient "plus" lamps (EE+) and keep standard ballasts.
4. Use 4 "Octron" (T8) lamps with electronic ballasts.
5. Replace 4 standard lamps by 2 Aurora IV lamps (or similar) and keep standard ballasts.
6. Install improved reflectors, remove 2 lamps, relocate the existing 2 lamps in the fixture and keep one standard ballast.
7. Install PRD (Edison 21) and keep the standard lamps and ballasts.
8. Replace two standard lamps with two Thrift/Mate lamps, e.g., one replacement per ballast circuit.

There could be other alternatives but time constraints prohibited considering all possibilities. To evaluate the 8 alternatives listed above, we have considered a life cycle approach. In this way, we consider every relevant cost item throughout the service life of a given alternative. In addition, we account for the time value of the money by using a discount rate. Also, we use the rated service life as the planning horizon for the cash flow analysis. Since bulb life and/or ballast life are not the same for different lamps and are dependent on the intensity of usage (whether one, two or three shifts), we use the Annual Worth method to evaluate alternatives with different service lifes. This method is based on the assumption that lighting systems will be replaced by identical models possessing the same cost.[4]

We have developed a generalized life-cycle cost profile for all the alternatives. This profile is depicted in Figure 20-1. We have assumed that change over will be carried out in an "as-lamps-burn-out" basis. Thus, the first cost is the increment "above" standard lamps and ballasts. It includes material and labor to install a given alternative. Similarly, any maintenance costs are included in an incremental basis. Cost figures have been obtained from trade literature and vendor quotations. Table 20-1 shows the total wattage per fixture and percent of energy savings for each alternative.

Table 20-2 depicts the lamp cost, the ballast cost and the total installed cost for each alternative. An alternative's incremental initial cost is the difference between its initial cost and alternative's 0 (do nothing) replacement cost. For alternatives that use existing standard lamps and ballasts, it has been assumed that they can be used for the remaining half of their service life.

Our economic analysis is rather conservative since it does not consider the air conditioning impact of lighting reduction. This could be justified whenever a building's conditioning savings (due to lighting reduction) are approximately the same as the additional cost of heating (during winter) required to make up for lighting heat reduction. This is generally the case in "thermally light buildings." But, in "thermally heavy buildings" and/or buildings with excessive lighting, additional savings can be realized from cooling load reduction.

Also, this study assumes that electricity prices and replacement cost for lamps and ballasts will remain constant throughout the planning horizon of each alternative.

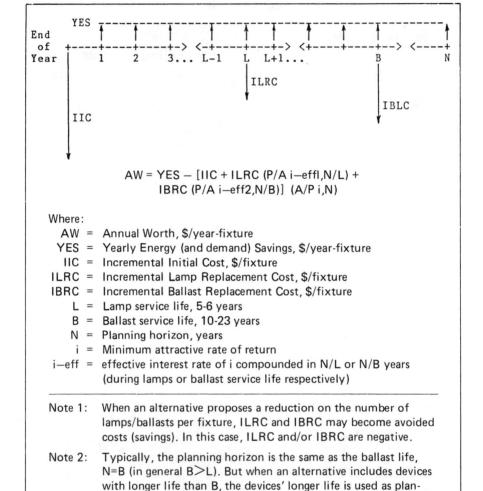

$$AW = YES - [IIC + ILRC \ (P/A \ i{-}effl,N/L) +$$
$$IBRC \ (P/A \ i{-}eff2,N/B)] \ (A/P \ i,N)$$

Where:

AW	=	Annual Worth, $/year-fixture
YES	=	Yearly Energy (and demand) Savings, $/year-fixture
IIC	=	Incremental Initial Cost, $/fixture
ILRC	=	Incremental Lamp Replacement Cost, $/fixture
IBRC	=	Incremental Ballast Replacement Cost, $/fixture
L	=	Lamp service life, 5-6 years
B	=	Ballast service life, 10-23 years
N	=	Planning horizon, years
i	=	Minimum attractive rate of return
i–eff	=	effective interest rate of i compounded in N/L or N/B years (during lamps or ballast service life respectively)

Note 1:   When an alternative proposes a reduction on the number of lamps/ballasts per fixture, ILRC and IBRC may become avoided costs (savings). In this case, ILRC and/or IBRC are negative.

Note 2:   Typically, the planning horizon is the same as the ballast life, N=B (in general B>L). But when an alternative includes devices with longer life than B, the devices' longer life is used as planning horizon. For instance, for Alternative 7 (PRD: Edison 21), N-21 years; and for Alternative 6 (reflectors), N=15.

Figure 20-1. Cash flow profile and annual work equation.

Table 20-1. Wattage per fixture and percent of energy savings.

ALTERNATIVE	Total Wattage* (watts)	Energy Savings (%)
0. Four Std. Lamps & 2 Ballasts (do nothing)	174	0.0
1. Four Energy Eff Lamps & 2 Std. Ballasts	155	10.9
2. Four Energy Eff Lamps & 2 Elect. Ballasts	119	31.6
3. Four Energy Eff+ Lamps & 2 Std. Ballasts	144	17.2
4. Four Octron Lamps & 1 Elec. Ballast**	106	39.0
5. Two Aurora IV Lamps & 1 Std. Ballast	92	47.1
6. Reflectors with two Std. Lamps & 1 Ballast	87	50.0
7. Edison 21 with 4 Std. Lamps & 2 Ballasts	122	30.0
8. Two Thrift/Mate Lamps, 2 Std. Lamps & Ballasts	117	33.0

*Source: "Fixture Comparison Data," (Sylvania EB-0-362)
**A four Octron lamps fixture needs one specifically designed electronic ballast instead of two.

Table 20-2. Lamp, ballast and installed cost for fluorescent lighting alternatives (2 x 4 ft. fluorescent fixture)

| ALTERNATIVE | COST | | | Installed |
	Lamp	Ballast ($/unit)	Other	Cost ($/fixt.)
0. Four Std. Lamps & 2 Ballasts (do nothing)	2.27	15	—	39.00
1. Four Energy Eff Lamps & 2 Std. Ballasts	3.15	15	—	42.60
2. Four Energy Eff Lamps & 2 Elect. Ballasts	3.15	23	—	58.60
3. Four Energy Eff+ Lamps & 2 Std. Ballasts	3.48	15	—	43.92
4. Four Octron Lamps & 1 Elec. Ballast**	3.67	39	—	53.68
5. Two Aurora IV Lamps & 1 Std. Ballast	7.00	15	—	29.00
6. Reflectors with two Std. Lamps & 1 Ballast	2.27	15	40.00	59.54
7. Edison 21 with 4 Std. Lamps & 2 Ballasts	2.27	15	30.00	69.00
8. Two Thrift/Mate Lamps, 2 Std. Lamps & Ballasts	2.27	15	11.34	57.22

*These figures include labor cost for installation.
**A four Octron lamps fixture needs one specifically designed electronic ballast instead of two.

Example

For Option 2, energy efficient lamps and electronic ballasts, the Annual Worth is:

$$AW = YES + [-IIC - IRC1 \ (P/A \ i-eff, 4) + IRC2 \ (P/F \ 10, 11)] \ (A/P \ 10, 23)$$

$$= 10.24 + [-19.52 - 3.52(1.39) + 30(.3505)] \ (0.1126)$$

$$= \$8.67/year$$

In this example, YES = $10.24 is the yearly energy savings; i–eff is the effective rate for a yearly rate i = 10%, compounded during a period equal to the lamp service life, L = 6 years. E.g., i − eff = (1 + i) **L − 1.[5] IIC is the incremental initial cost over the standard option. IRC1 and IRC2 are the incremental replacement costs, for lamps and for ballasts respectively.

The savings due to energy and power reduction are computed in terms of a weighted average demand charge ($6.05/kW-month) according to Oklahoma Gas & Electricity's PL-1 schedule (service level 5). Consumption charge is $0.03639/kWh.[6]

Most of the state office buildings operate during 8 to 10 hours per day, but are occupied 2 to 4 additional hours by custodial personnel. Hence, we consider 12 hours of operation per day. In addition we have assumed a 10% interest as a minimum rate of return for each alternative's cash flow.

Annual Worth (equivalent net dollar savings per year) is the measure of merit for each alternative. Note, this means a higher number is preferred over a lower number if everything else (other valuation factors as lumens/watt, color rendition, building occupants' opinion, etc.) is equal. However, everything else is not equal, and a particular building, or even room application may require a specific alternative. But, in general OPA will select the two or three "best" alternatives to simplify purchasing and replacement. Annual Worth results are listed in Table 20-3.

LIGHTING EXPERIMENT IN THE DEPARTMENT OF TRANSPORTATION BUILDING, OKLAHOMA CITY

To gain insight on the actual performance of fluorescent lighting systems available to the State Buildings, we asked vendors to install several fluorescent lighting options in a large drafting room of the Department of Transportation at the State Capitol Complex, Oklahoma City. Due to time constraints and sample availability, we were able to test only a limited number of options. The selected room had a number of different circuits with fourteen four-40w-lamp fixtures each (2 x 4 ft).

Table 20-3. Annual worth for fluorescent lighting alternatives
(Equivalent net annual savings)

Alternative	Annual Worth ($)
1. Energy Eff Lamps & Std. Ballast	2.66
2. Energy Eff Lamps & Elect. Ballast	8.67
3. Energy Eff+ Lamps & Std. Ballast	4.37
4. Octron & Elec. Ballast	11.31
5. Aurora IV Lamp & Std. ballast	16.34
6. Reflectors, Std. Lamp & ballast	14.79
7. Edison 21, Std. Lamp & ballast (30% savings)	8.00
8. Thrift/Mate, Std. Lamp & ballast (33% savings)	6.16

Table 20-4 lists the results of the experiment on options available from local vendors at the time we started the study. This table also includes average Amps and kW readings. In addition, after measuring during a period of time the kWh consumed by each circuit, we computed the relative energy consumption of each one of the options with respect to the current situation (standard lamps and ballasts). Measured line voltage at the breaker panel was about 270 volts.

Finally, we measured the average illumination level under each circuit's area. For this purpose footcandle readings were taken for each individual circuit (the other circuits in the same room were turned off). The readings were taken during night at "drawing board height" (about 3 ft above the floor and 9 ft below the fixture level). The footcandle figures in Table 20-4 are mean values and have certain bias due to light reflected from partitions and walls.

CONCLUSIONS

When using Annual Worth as the evaluation criteria, the Aurora IV, or equivalent (improved-color-rendition-with-higher-lumen lamps), appear to be the best. However, we need much

**Table 20-4. Results of fluorescent lighting experiment in
the Department of Transportation building**

No. of Circuit	Option	Current (Amps)	Power (kW)	Energy Index	Illum. (fc)*
19	Standard lamps & ballasts (40w)	9.0	2.40	1.00	103
14	Energy eff lamps elec. ballasts (36w)	4.8	1.31	0.57	65
23	Reflector, half std lamps & ballasts	4.7	1.20	0.52	92
17	Power reducer, 33%, std lamps & ballasts**	5.5	1.54	0.67	82
16	Improved color rend lamps & std ballasts**	4.2	1.23	0.53	91
22	Octron Lamps with elec. ballast***	5.5	1.54	0.70	152

*A footcandle (fc) is the illumination on a surface one square foot in area on which there is a uniformly distributed flux of one lumen.[7]
**All options, except these two systems—with 2 lamps per fixture, had 2 lamps/ fixture.
***This option requires one especially designed electronic ballast per fixture.

more experimentation with this new kind of lamp. We will use them on an experimental basis in some of the smaller buildings with current excessive illumination. For other buildings, the selection will be improved reflectors. The replacement lamps will be energy efficient and/or Octron lamps (with electronic ballasts). The specific recommendation will depend on the actual illumination levels and the requirements of a given building.

By the time these recommendations are fully implemented, significant savings will be realized by each one of the state buildings in the Capitol Complex. The savings can be from 30 to 50% of the actual lighting energy cost. Further cost reduction can be attained, since other energy cost reduction measures are being recommended on heating, ventilation and air conditioning systems. Nevertheless, savings have started to show up in all those buildings that are carrying out our energy cost reduction recommendations. As a part of an ongoing energy management program, Office of Public Affairs will make available the results of this and other studies, being carried out by the Oklahoma State University team, to other Oklahoma State agencies.

ACKNOWLEDGEMENTS

The Energy Division of the Office of Public Affairs and the research team from Oklahoma State University are grateful to the firms and individuals that supplied cost and technical information, and/or participated in the lighting experiment. We also express our appreciation to building management personnel, especially from the Department of Transportation Building, for their valuable cooperation.

References

[1] Edison 21 Fluorescent Monitor, National Energy Research Corporation, brochure NE-101, Newport Beach, California 92663

[2] Thrift/Mate Fluorescent Lamps, Sylvania Industrial/Commercial Lighting, GTE, brochure FL-803, Danver, Mass. 01923

[3] Silverlux Fluorescent Reflectors Turn On Energy Savings, Energy Control Products, 3M Center, St. Paul, MN 55144-1000.

[4] Riggs, J.L. and T.M. West: *Engineering Economics,* 3rd ed., McGraw-Hill, New York, 1986.

[5] Oklahoma Gas and Electric Company, *Standard Rate Schedule,* PL-1 (service level 5), effective on and after July 2, 1987. Oklahoma City, OK 73101-321.

[6] White, J.A., M.H. Agee, and K.E. Case: *Principles of Engineering Economic Analysis,* 2d ed., Wiley, New York, 1984.

[7] Illuminating Engineering Society: *IES Lighting Handbook,* 5th ed., New York, 1972.

Chapter 21
Sacramento Municipal Utility District Commercial Lamp Installation Program

Barbara Coulam

INTRODUCTION

The Sacramento Municipal Utility District, a customer-owned utility, is governed by a board of directors. It is a large-scale electric generation, transmission, and distribution utility. The District serves 890 square miles, including 398,000 residential and commercial customers. Of these, 18,000 are commercial customers with less than 30 kW demand and 4,000 with 30-199 kW demand (Figure 21-1).

Electric power is generated through nuclear (930 MW), hydro (660 MW), geothermal steam (72 MW), photovoltaic (2 MW), and purchased power (360 MW).

SMUD included the Commercial Lamp Installation Program (CLIP) in the 1986 budget as a pilot program. CLIP was designed to test the cost effectiveness of installing energy-saving fluorescent lamps. Also, almost 80 percent of those businesses were operated in leased facilities and therefore not inclined to implement the recommendations (Figure 21-2).

In May 1986, the Western Area Power Administration agreed to fund $30,000 to the Project for installation labor costs.

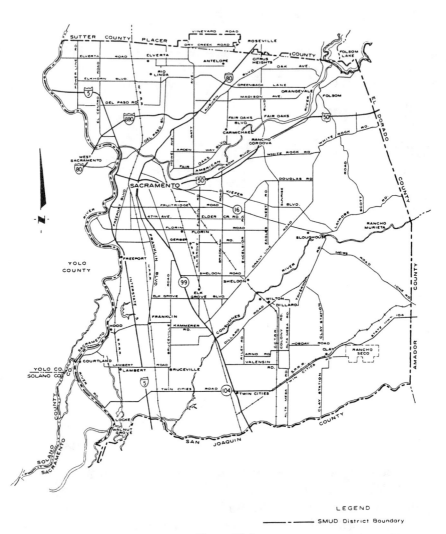

LEGEND

———— · ———— SMUD District Boundary

Figure 21-1

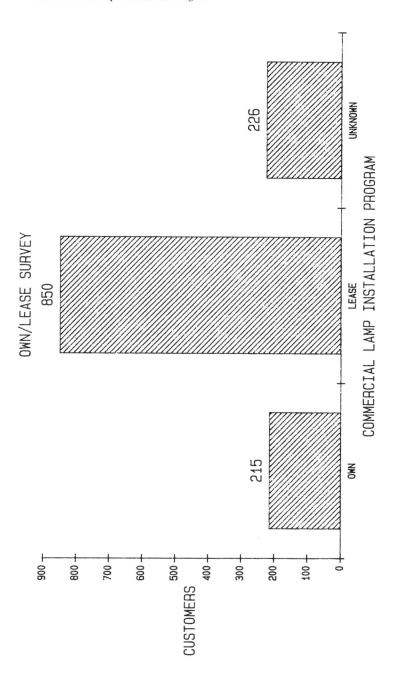

OWN/LEASE SURVEY

Figure 21-2

PILOT PROGRAM

CLIP began operations July 14, 1986, with CACS trained auditors soliciting customers to participate in the program. Installations began August 4, 1986. By the end of the initial pilot program, December 31, 1986, the program had installed a total of 54,362 F40 and F96 energy-saving fluorescent lamps in 1,278 small commercial facilities in the SMUD area. These installations represented a summer peak load reduction of 478 kW and 1,321,257 kWh in energy savings. The program was 42 percent over goal for installations and 62 percent over goal for reduced kW.

GOAL

The CLIP program goal was to determine if a program to install energy-saver fluorescent lamps at no charge to the customer could effectively provide load reductions and savings from small commercial customers.

OBJECTIVES

The objectives of the program were:

- To implement the energy-saving fluorescent lamp recommendations identified by the small commercial audit program,
- To develop cost benefit information
- To determine customer acceptance
- To determine if direct lamp installation increased implementations from small commercial audits,
- To determine vendor and lighting maintenance contractor acceptance, and
- To determine the feasibility of extending the program beyond the 1986 pilot test program.

COSTS

The total costs for the pilot program July 14 through December 31, 1986 are as follows:

Labor (contract)	$ 85,804
Labor (staff)	30,000
Incentives (lamps)	72,635
Equipment	1,300
Vehicles	6,000
Material replacement	828 (Exceptions)
TOTAL COSTS	$196,567

COST BENEFIT RATIO

Using the Barakat, Howard & Chamberlin, Inc. Demand-Side Management Program, two scenarios were developed. Under the first scenario a 3-year program was assumed, a 4-year lamp life, with a 25 percent impact persistance through the year 2015, and with marginal costs in 1986. The second scenario was the same as the first but without the 25 percent persistance. Using the following ranges of cost benefit ratios, SMUD made the decision to continue CLIP as an operational program:

	I	II
• All Rate Payers	3.08	2.55
• IRL	1.09	.94
• Participant	6.04	10.35
• Utility	5.69	3.34

PROGRAM OPERATIONS

Customer Size

CLIP was designed to install energy-saving fluorescent lamps in place of standard fluorescent lamps in small commercial customers with less than 30 kW demand at no charge to the customer. This customer typically uses less than 48,000 kWh annually.

Proposed Number of Installations

As a pilot program, CLIP targeted the installations of energy-saving fluorescent lamps in 900 small commercial customer businesses by December 31, 1986. Six hundred facilities were to be installed for customers who have previously received a Commercial Apartment Conservation Service (CACS) audit or who had refused an audit. The remaining 300 installations were to be made for accounts that had not been previously solicited for audits.

Maximum Number of Lamps

Under CLIP, a maximum number of lamps was installed: up to 100 F40 lamps or 50 F96 lamps or a combination to equal approximately .75 kW per facility could be installed. There was no minimum. The lamps were purchased under a competitively bid California State Contract.

Solicitations

CLIP customer lists were developed from the billing master and were sorted by zip code and street address. CACS solicitations had used the same lists and zip codes. Door-to-door solicitations within these areas were preceded by an introductory letter accompanied by a program brochure and offer plus a group relamping brochure. The letter was followed 5 to 10 days by the auditors visiting the facility.

Three CACS trained auditors preinspected and evaluated the various facilities. Three teams, two installers each, followed several days later making the lamp installations.

Solicitation Results

Fifty-seven percent of the 2,250 customers contacted accepted an installation (Figure 21-3). Of the customers who had the lamps installed:

- 28% had previously had a CACS audit,
- 18% requested an installation (offer by bill message or referral),
- 9% had refused an audit but accepted an installation, and
- 44% were solicited by an introductory letter preceding a cold canvas.

Pre-Inspection

A pre-inspection of a facility must show the following to be eligible:

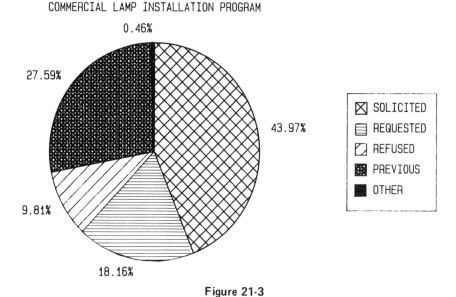

Figure 21-3

- Must be a small commercial, non-demand account (less than 30 kW),
- Lights must operate across the utility peak,
- Does not already have energy-saving fluorescent lamps,
- Lamps must be in conditioned space, 12 feet or less from the floor and readily accessible,
- Ballasts must be compatible with energy-saving lamps,
- Customer understands that previously delamped fixtures are not eligible,
- Lamps must be installed within program operating hours,
- Customer understands that inoperable fixtures, fixtures with ballasts in process of failing, obvious mechanical problems, or F40 single-pin lamps are not eligible,
- Customer does not have existing lighting maintenance contract,
- Customer is informed that the removed lamps will be disabled,

- Customer is informed that extra lamps will be eligible for an incentive of 40% rebate under a separate incentive program.

Lamp Installtions

The installations were made by the two-person installing crews. Field time was optimized by grouping installations within a specific zip code area, scheduling appointments, and "wild cards" to back fill extra field time.

A packet of information was left with the customer after the installation. It included a "thank you for participation" letter, vendor information, a customer survey and educational material on the benefits of energy-saving fluorescent lamps.

At the time of the lamp installations, a sticker was placed in the fixture. The sticker is designed to encourage the customer to relamp with the energy-saving fluorescents when necessary. It also gives the program some control if premature lamp failures should become a problem in the future.

Installed Lamps by Type

A total of 54,362 lamps were installed during the pilot program. Energy-saving fluorescent F40cw (38,646 lamps), F96cw (12,941 lamps), F40ww (1,035 lamps), and F40lw (1,710 lamps) were installed (Figure 21-4).

Post Inspections

Twenty percent of the installations were post inspected by staff and auditors to verify installation and customer attitude. Midway through the pilot period, the post-inspection rate was reduced to 10 percent due to the lack of problems.

Non-Participation (Figure 21-5)

The reasons for customer non-participation were classified as follows:

A. Existing energy-saving lamps – 6.6%

B. Customer was not interested – 10.4%

C. Ceilings not accessible - 2.1%

D. Color of desired lamps not available under program – 3.1%

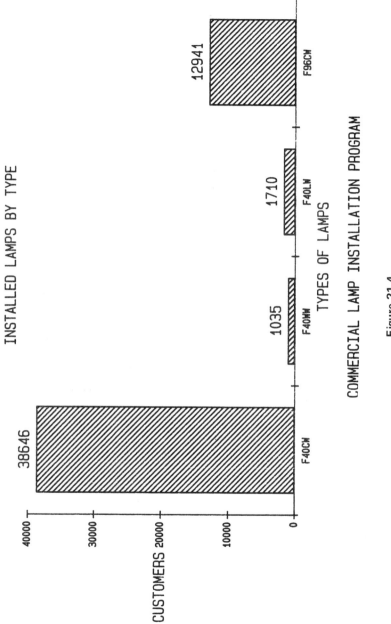

Figure 21-4

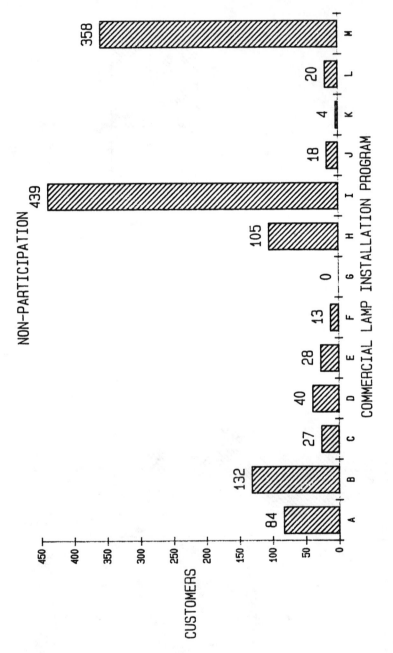

Figure 21-5

E. Incompatible system (old) — 2.2%

F. Operating hours not across peak — 1.0%

G. Lack in information — 0%

H. Exterior lamps or unconditioned space — 8.3%

I. Decisionmaker not available - 34.6%

J. Inconvenient time — 1.4%

K. Concerned about ballast and lamp failure — .32%

L. Customer had a lighting maintenance contract — 1.6%

M. Other (usually non-English-speaking) 28.2%

In the current program, staff is investigating methods to target the "decisionmaker not available" problem.

Exceptions

BALLASTS. Program staff anticipated premature ballast and lamp failure. The problem has failed to materialize. The following "exceptions" costs have been included in the cost of the program:

- Premature ballast failure — 26 failed ballasts (replaced)
- Premature lamp failure — 25 failed lamps (replaced)
- Lamps not properly installed — 15
- Damage to customer facilities — 1

Six additional ballasts were shorted out during pre-inspection or installation and were included in the costs. They occurred when the customer was not willing to turn off the lights in the bi-pin F40 lamps. As a result, willingness to turn off the lamps during pre-inspection and installation has become an eligibility requirement.

DIFFUSERS. Old, brittle diffusers were a slight problem. They are very fragile and break or crack easily. Originally, the program was replacing them if broken during installation. The customer is now asked to accept the responsibility during the pre-inspection and installation unless the damage is accidentally done during installation and then the program accepts the responsibility. Damage to customer facilities has been minimal.

Approximately three cents per installed lamp can be added

to the cost of the program, including ballasts, lamps, customer property, and diffuser replacement costs.

Record Keeping

A data base was established to capture demographics, solicitations, installations, installed lamps, installed kW and kWh, and ineligible customers.

A manual system was used for recording exceptions. Lotus 123 spread sheets were used for inventory and budget control, labor hours, and costs, etc.

MAJOR CONCERNS/PROBLEMS

A lack of storage space for a month of backup lamp inventory and erratic lamp vendor deliveries presented some problems. Warehouse procedures have been changed to accommodate at *least* a month of backup inventory. The vendor has not been cooperative.

Procedures for ordering the lamps were modified to accommodate program needs and vendor shipping problems. There are still some problems in this area and they appear to not have an immediate solution.

TRANSFERABILITY

The Commercial Lamp Installation Program is very transferable to other utilities. The concepts of CLIP (the immediate installation of kW-reducing products) has a high degree of potential if the following criteria are considered:

- The utility is a summer or winter peaking utility and reduction of kW during business hours is important.

- The utility can benefit from energy savings.

- The utility has an energy audit, pre-inspection/inspection, or other program requiring commercial customer contact.

- The utility marginal costs create a positive cost effective climate for the development of a program such as CLIP,

- The utility is able to purchase the lamps at a favorable price and appropriate vehicles are available to carry lamps,

- The utility needs positive customer public relations.

SUMMARY

CLIP program staff anticipated that each crew would be able to make three installations per working day. Each installation was to be approximately 50 F40 energy-saving fluorescent lamps representing an installed .3 kW savings per job.

In operation, the program installed an average of 42 lamps, .31 kW per job and 3.55 jobs per day per crew. The program ran 42 percent over goal for installations and 62 percent over goal for reduced kW demand during the pilot phase. Essentially, the same results are being obtained in the operational 1987 program with minor variations for demographics.

The customer attitude survey received a 35 percent response rate. The survey was designed to pick up both positive and negative attitudes from the customer. There has been an overwhelming positive response rate. Less than two percent had negative comments. It is difficult to measure the positive public relations that CLIP has achieved.

CLIP was funded in 1987 Energy Services budget and is considered operational.

Chapter 22
Lighting Projects
Case Studies

CASE STUDY 22-1
TESTING LOWER LIGHTING LEVELS
AT WESTERN MARYLAND COLLEGE

Western Maryland, Westminster, MD, an innovative college, kept with that tradition, when their operations department was the first to test Honeywell's fluorescent lighting control system.

Prototype control and output modules of the Honeywell system were installed in the college cafeteria in early December 1986. The test has been successful, and control systems will be installed in at least 10 campus buildings. At first, only lamps that are on 14 to 16 hours a day will be controlled.

"Many people take lighting levels for granted," said Edgar Sell, Jr., director of physical plant for the college. "They don't realize it consumes a lot of energy." But Sell did realize the energy costs of lighting and the difficulty in reducing light levels to save energy. "Buildings are usually overlit, due to engineering and design considerations," he said. "If later we want less light, we can't adjust it. The gym, for example, was designed for 4 watts per sq ft. Current requirements are for 2 watts per sq ft.

"Disconnecting tubes is not satisfactory because it is labor-intensive. Removing lights is ineffective; it makes lighting uneven. We can't install energy-efficient lamps with ordinary ballasts, because the ballasts would burn out. If lights are turned on and off too frequently, the tubes burn out much faster."

The fluorescent lighting control system developed by Honeywell's Building Controls Division gives them more flexibility. "We will be able to vary the output intensity of fluorescent lights," said Sell. The cafeteria was selected for the test because it is a large room with both windows and fluorescent lights. Sell was concerned that the control system might affect lamp life.

So he replaced all ballasts and tubes in one section of lights while leaving existing ballasts and tubes in the other. During the first four months of the test, no tubes or ballasts burned out in either strip of lights.

Sell wanted uniform wattage in all fixtures as well as uniform lumination. He followed Honeywell's recommendation to install new lamps and clean the fixtures. The control system dimmed the new lamps to the light level of the old lamps, but used 30 percent less electricity.

Testing of the demand-level dimming control was conducted during the lunch period on an overcast, cloudy day. The lamps were slowly dimmed, and the students continued their activities without looking up or noticing that the light levels had been reduced. This experiment was conducted in order to eliminate concerns that the students would notice the slow dimming.

Fluorescent lights that will be controlled include those in hallways, the gym, classrooms, bookstore, post office, library and kitchen. No payback or dollar savings were calculated for the test installation. Additional systems will be installed throughout the year.

CASE STUDY 22-2
SPORTS COMPLEX RETROFIT BRINGS
ELECTRIC AND GAS BILL SAVINGS

Application of state-of-the-art techniques at the Rolling Meadows Sports Complex, Rolling Meadows, IL, is saving more than $50,000 per year on electricity and gas bills, according to Rudolph Nelson, director of parks and recreation.

Designed and installed by Air Comfort Corporation, Broadview, IL, the updated systems have achieved a 34 percent reduction in the cost of electricity and 46 percent on natural gas, he said.

In 1982, electric and gas costs at the complex totaled $103,505 and $43,970, respectively. The utility services supplied the needs of a 200-seat ice arena, a warming room, a gymnasium, an outdoor swimming pool and showers, as well as baseball facilities. With costs rising at 20 percent annually, Nelson initiated a study by Air Comfort Corporation, which designs energy-saving retrofit systems. Michael E. McDonough, chief engineer, headed the Air Comfort team in charge of the modernization project.

After an intensive study of existing facilities, Air Comfort recommended a four-phase solution to the problem: installation of high-intensity, low-consumption lighting equipment; the addition of stand-by generating equipment to implement adoption of a more favorable electric rate schedule; improved water heating facilities, and a new reclaim system to save heat formerly wasted in the skating rink ice-making process for a number of purposes throughout the complex.

Substantial savings were accomplished through a complete redesign of the lighting systems. Energy-wasting incandescent wallwash fixtures in the corridors and meeting rooms were replaced with high-efficiency fluorescent units, resulting in a 60 percent reduction in electric costs. Incandescent outdoor lighting fixtures were ruled out in favor of energy-efficient, high-pressure sodium-vapor units, saving 62 percent.

Mercury-vapor and fluorescent fixtures in the gymnasium were supplanted by high-pressure, sodium-vapor units, stepping up illumination intensity 50 percent and cutting operating costs 75 percent. Fluorescent wallwash units in the skater's warming room gave way to fluorescent ceiling fixtures which doubled lighting intensity, but cut costs 65 percent. In the ice arena, mercury-vapor and indirect fluorescent equipment was abandoned and replaced with metal halide ceiling fixtures with a 70 percent reduction in operating cost.

With rising demand and energy charges, Air Comfort recommended mechanical changes that permitted the application of a lower rate schedule. These revisions also eliminated the threat of power failure that confronted the complex—a threat that could spell disaster to the ice rink on a hot summer day.

A 305-kW, diesel-powered generator was installed in its own outdoor concrete block building. Replete with a 1,000-gallon oil storage tank, it also has an 800-ampere transfer switchbox adequate to handle the electrical demands of the entire complex. This diesel will limit power failure impact to 15 seconds, the time it takes for the automatic start-up of the generator.

It also paved the way for adoption of a much lower utility rate for optional, interruptible service.

To recover heat formerly wasted in the ice-making process for the skating rink, Air Comfort installed a reclamation system which harnesses the heat and holds it in an existing water storage tank. Now continuously heated and recycled through the tank, water formerly heated by a gas boiler is provided at no cost. The reclamation technique affords free potable hot water and heats the skating rink and spectator areas with a 42 percent reduction in fuel cost.

The Rolling Meadows-Air Comfort contract included a guarantee that the entire retrofit cost would be repaid in one year in energy savings or Air Comfort would make up the difference. The commitment was met within the presceibed time limit, according to Air Comfort.

CASE STUDY 23-3
PRIZE-WINNING LIGHTING DESIGN BRIGHTENS
MISSOURI SUPREME COURT LIBRARY

When McMichael Auman Consultants, St. Louis, MO, and the Christner Partnership completed the lighting design for the Missouri Supreme Court Library, the power density was reduced from 4.0 to 2.8 watts per sq ft. The French Renaissance architectural design flourishes were enhanced, emergency egress lighting was provided where none existed before and the project was completed within budget.

The work was accomplished without extensive rewiring or exceeding an extremely tight construction budget allowance of $5.50 per sq ft for the 10,000 sq ft library.

The focus of the design work was to provide aesthetically pleasing and more effective lighting in the circulation, task and stack areas without interrupting library operations. The library, built in 1906, is located in Jefferson City, MO.

McMichael Auman Consultants is a St. Louis-based firm specializing in the design of engineered building systems for lighting, heating, air conditioning, plumbing, fire protection, security, communications, and power generation and distribution. Harry Auman, principal of McMichael Auman, and George W. Johannes, managing partner of Christner, led the lighting renovation design team.

The Illuminating Engineering Society of North America has awarded its Edwin F. Guth Award of Merit to the two firms for lighting design excellence at the Missouri Supreme Court Library.

Although the library was historically renovated several years ago to its original French Renaissance style, the lighting system had not been included in the renovation. What remained was a row of 1950-vintage, 5-foot-square, brass and copper chandeliers that housed fluorescent and incandescent lamps. They provided only 15 footcandles of ineffective illumination and visually obscured the beauty of the restored two-story-high coffered ceiling.

In addition, strip fluorescent lighting fixtures in the two-level stack area were a source of glaring illumination for those searching for books. The stacks border the room lengthwise with the second level of stacks serving as a balcony overlooking the primary circulation area. The two levels of stacks are separated by a translucent glass floor.

"By removing those bulky chandeliers, we were able to open up the view of the ornate ceiling," recounts Johannes. "We then decided to highlight the ceiling by using indirect lighting sources based on top of the second floor stacks."

The fixtures used in the indirect lighting are housed in simple black boxes containing two lamps each, one 400-watt metal halide and one 250-watt high pressure sodium bulb. The lamps generate 15 footcandles of light on the second floor stacks and 20 footcandles of ambient lighting for general areas on the main floor.

Harry Auman combined the high pressure sodium lamps with their warm amber color and the metal halide lamps with their bluish white light for a natural lighting effect in the room.

"We were concerned with performance, glare, the budget and the historical accuracy of any visible fixtures," explains Auman. "Although the indirect fixtures are not historically accurate, they are not visible to library users. It is hard to say what fixtures would have been historically accurate since the library originally depended heavily on daylight."

To provide adequate task lighting, Johannes suggested an idea for tabletop lamps from Henri Labrouste's Bibliotheque National (1862) in Paris. Auman found what Johannes envisioned, but the fixtures were designed as ceiling-mounted units. Auman worked with the manufacturer to obtain the needed modifications which involved turning the fixtures upside down, reducing their height from 36 to 24 inches and adapting the brass bases for permanent tabletop mounting. The four lamps on each fixture extend from the center brass piece at right angles from one another, defining specific task areas at every table.

New illumination was needed to replace old fluorescent lamps in the two-story stack area with its glass ceilings and floors. The new system was designed almost by accident when Johannes and Auman were testing a low-profile fluorescent fixture for use on both stack levels. As they were measuring the fixture's footcandles on the first level, a Missouri Supreme Court justice observing the men's work suggested holding the unit upside down. The justice wanted to see if the fixture could illuminate the second floor stacks through the glass ceiling of the first level. Johannes and Auman modified the fixture's design by removing the sheet metal above the lamp compartment, allowing light from the fixtures to shine upward through the glass ceiling. The modification provided 15 to 45 footcandles of light for the stacks, minimized glare, increased head room and cut in half the number of fixtures originally needed for both levels of stacks.

The final improvement to the library's lighting system included the addition of emergency egress lighting. "We put some battery-inverter units in strategic areas of the stacks where there were none before, bringing emergency lighting up to code," Auman noted. Refer to Figure 22-1.

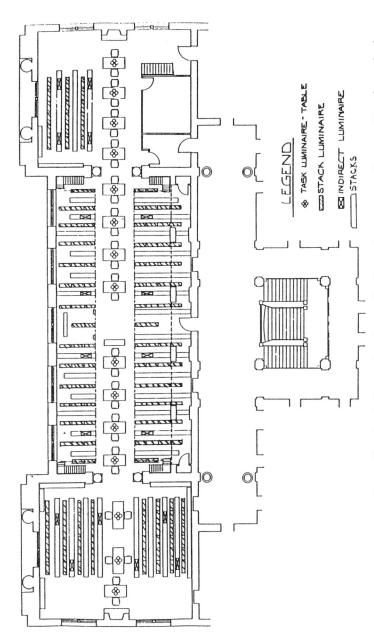

Missouri Supreme Court Library first level floor plan shows the library's narrow stacks extending from the longer north and south walls creating a linear east-west task area. Each table in the task area is illuminated by a four-lamp, green shaded fixture. Shorter rows of stacks on the balconied second level are located directly above the first-floor stacks. A light dotted line marks the edge of the balcony's glass floor. The darker dotted line denotes the edge of a seldom used third-level balcony.

—Floor plan provided by The Christner Partnership, Inc.

Figure 22-1. Lighting plan for Missouri Supreme Court Library.

LEGEND

⊗ TASK LUMINAIRE - TABLE

▨ STACK LUMINAIRE

⊠ INDIRECT LUMINAIRE

▭ STACKS

CASE STUDY 22-4
LUMINAIRE CONTROL SYSTEM OPERATES
WAREHOUSE LIGHTING BY RADIO SIGNALS

A state-of-the-art lighting system was added to Affiliated Food's other modern warehouse features such as electronic ordering, five miles of the most modern type of conveyor system, and laser scanners to sort coded boxes.

Luther Batie, manager of maintenance with Affiliated Food Stores, Inc., chose a General Electric lighting system for the dry grocery portion of their recently built Keller, TX, distribution warehouse. Lighting in the 450,000 sq ft area is provided by REM® Controls for Luminaires, which selectively control 300 GE Lowmount® II luminaires, automatically varying their wattage or turning them off according to the activity in the area, thus saving energy costs.

The dry grocery portion of the warehouse handles case sales of nonperishable food and grocery products. "Half a million cases a week move out," says Batie. "We sort 7,000 to 8,000 cases per hour from palletized racks, and our inventory of $25 million turns over every two weeks." The REM system is programmed so that half the luminaires turn off automatically during the hours when goods are not being moved in or out of an area. Each Lowmount II luminaire houses a 400-watt GE Lucalox® high pressure sodium (HPS) lamp. Frequency-modulated (FM) radio signals transmit commands to the luminaires, telling each one to turn off or to operate at any one of seven wattage levels, from 150 to 450 watts in 50-watt increments.

"The REM system is really versatile," says Jerry Couch, Supervisor of Electronic Maintenance. "It operates by radio signals, without any hard wiring, so we can move fixtures around or reprogram them without worrying about control rewiring."

REM Controls for Luminaires can selectively control any number of high pressure sodium luminaires in up to 252 "zone addresses" or sets of luminaires. Luminaires are individually programmed, so they need not be physically near each other to be programmed to receive the same signal.

The REM transmitter sends out a computer set of instructions approximately every 10 seconds, operating seven days a week, 24 hours a day. There is no need to reprogram the REM controller unless changes in schedule are required.

The Affiliated Food Stores warehouse is divided into about 55 zones, with these combined into 13 groups. Groups simplify programming, and are formed from any combination of zones. "We have most of the fixtures programmed to operate at 250 watts, supplying 25 footcandles at aisle level," explains Couch. "We've found this to be the most energy-efficient for us, and it is usually enough because we have skylights providing additional light. On cloudy days, I manually override the system and raise the wattage to 400 watts and the light level to 50 footcandles."

Because HPS lamps take a few moments to warm up, luminaires at intersections of aisles are equipped with automatically switched quartz instant/on safety lighting. These lamps come on immediately when power is restored after a momentary outage.

Vertical Footcandles an Important Consideration

The 300 Lowmount II luminaires are mounted at 27 feet over the aisle floor, midway between storage racks. They are spaced at 30 feet along each aisle. The prismatic refractor on the luminaires spreads out the light to provide high light levels on vertical surfaces, thus making it easier for the fork lift operators to read letters on the boxes.

"There are no shadows in the aisles," comments Couch. "You don't have the spotlight effect you have with some light sources, because of the diffusers."

Lighting System Easily Maintained

Couch estimates that they would have needed at least 5,000 fluorescent lamps to get the present light level compared to the 300 HPS lamps. Fluorescent tubes have a much shorter life than HPS, too, so we'd need a full-time person to change light bulbs, he said. According to Couch, the luminaires operate 7,500 hours per year. The rated life of HPS lamps is 24,000 hours plus, so they last over 3 years in the Affiliated Foods warehouse.

The loop, cord, and plug mounting assembly is easily disconnected if Couch wants to reposition a luminaire. Each luminaire is marked with its zone number, so if it is taken down for replacement or maintenance, a spare can be programmed as it is put up. "It takes five minutes to change out a fixture," according to Couch. "We don't even have to turn the power off."

Maintenance is simplified, too, he said because the luminaires contain an activated charcoal filter that traps airborne contaminants before they can enter the optical assembly. This helps keep the inside of the luminaires clean, avoiding the deterioration of light output that could result from dirt build-up. "The fixtures don't get very dirty," says Couch, "even though we fog every week against insects."

In addition to the REM system, lighting in the Affiliated Food Stores warehouse includes 150 GE Lowmount 400 luminaires, utilizing 400-watt Lucalox HPS lamps. A row of these luminaires provides 50 footcandles along each side of the dry grocery area: one at the outlet of the conveyor system and the other where trains are loaded or unloaded. Lowmount 400 luminaires are also used in freezer areas.

The consultant for the entire warehouse project was Food Plant Engineering, of Yakima, Washington. The 277-volt lighting system was installed by General Engineering of Fort Worth, Texas. John Rice, GE Lighting Systems Engineer, Dallas, supplied photometric information and expertise on the REM Controls for Luminaires.

CASE STUDY 22-5[1]
RETROFIT CASE STUDIES
CHILDREN'S HOSPITAL, ST. LOUIS, MO.

Children's Hospital in St. Louis, Missouri, is a 3-year-old, 12-story, state-of-the-art building of approximately 500,000 square feet. The lighting requirements were common to those of a hospital, and most of the retrofit performed took place in non-patient areas such as corridors, offices, nurses stations, etc.

---

[1]"Upgrading An Existing Lighting System: The Lighting Maintenance Perspective," Cary Mendelsohn.

It turned out to be one of the largest retrofits in St. Louis. It involved 3643 fixtures. In every 2 x 4 four-lamp fixture (2,592) a custom designed Magnalite reflector was installed, one ballast was disconnected and 2 new F40/WW lamps were installed. In any two-lamp, 4-foot fixture (either 1 x 4 or 2 x 4) a reflector was installed, as well as a 50% current limiter and two new F40/WW lamps.

A pilot program of 100-200 fixtures was run to make sure the customer was satisfied. Light meter readings showed a *gain* in footcandle measurement. The job was performed by my company, Imperial Lighting Maintenance Company, in December of 1987 and January 1988. The projected energy savings per year were $83,076. Based on a cost of $114,000 for the job (which included new lamps) the hospital realized a 1.3-year payback with a return on investment of 72.%.

CASE STUDY 22-6
CHICAGO TRIBUNE, CHICAGO, IL.

This is a 5½-year-old building with three major types of fluorescent lighting fixtures: 2 x 4 four-lamp fixtures, two-lamp 8-foot slimline industrial fixtures, and two-lamp 8-foot High Output industrial fixtures.

The newspaper investigated all types of retrofit possibilities such as current limiters, reflectors, energy-saving lamps, electronic ballasts, new fixtures, and finally settled on the "No-Watt" current limiters, manufactured by Remtec Systems of Duarte, Ca., and sold to *The Tribune* by Energy Command Systems of Fairbury, Il.

The key elements in this job were 1) that *The Tribune* had originally designed excess lighting into the system; and 2) the fixtures were not being relamped on a regular cycle, therefore the light output of the existing lamps had deteriorated. Even though a 30% current limiter was installed along with new lamps, light levels increased.

With regard to specifics of this job there were 1850 2 x 4 fixtures in which two 30% current limiters were installed as well as the fixture being washed and relamped. There were 1180 two-lamp, 8-foot slimline and 420 two-lamp, 8-foot High Output fixtures in which a 30% current limiter was installed and the fixture was washed and relamped.

The Tribune experienced a savings of approximately $72,334. The cost of the job was $110,897. Payback therefore was 1.53 years and there was a return on investment of 65%. One offshoot to this job was the fact that the current limiter allowed the ballast to run cooler, thereby extending the life of the ballast. Keep in mind also that the total cost of the job included the cost of all new lamps plus the labor to wash the fixture and install the lamps. This is a cost which would have had to have been borne in any case.

CASE STUDY 22-7[1]

Using the standard energy-saving T-12 (1½-inch diameter) fluorescent lamp with standard or energy-saving ballast systems has reduced light output when operated in the same luminaire. The T-8 (1-inch diameter) fluorescent lamp, however, is an energy-efficient, fluorescent lighting system that delivers the same light output from the luminaires as a standard F40T12 lamp/ballast combination, while reducing overall energy requirements and providing good color rendering at color temperatures of 3100, 3500 and 4100K.

The first T-8 lamp was a 4-foot, 32-watt unit. Because of optical and thermal improvements when operated in a luminaire as well as improved phosphor technology, the T-8 system is able to provide the same amount of light as a 40-watt, T-12 system. The T-8 lamp uses medium bi-pin bases and is physically compatible with luminaires designed for 4-foot, T-12 lamps. However, a new ballast is required for the T-8 lighting system since it operates at 265ma as opposed to 430ma for T-12 lamps. Lamp lumen depreciation of the T-8 system is also increased to 90% as compared to 88% for a T-12 system because of the reduced current and the use of rare earth phosphors.

---

[1]Lighting Retrofit Opportunities Make Good Business Sense, R. Arnold Tucker, P.E., C.E.M.

In renovating its office and manufacturing facility in Waltham, MA, Hewlett Packard put a high priority on the type of lighting selected. Efficiency and return on investment received the greatest attention. The new lamps and ballasts in the existing fixtures are T-8 fluorescent lamps and electronic ballasts throughout the approximately 460,000-square-foot facility.

Compared to the standard T-12, 4-foot fluorescent lamps, the T-8 fluorescent systems make possible significant energy savings of up to 46%. This combination provided the most efficient 4-foot system available anywhere at any cost. With the highest fluorescent lamp efficiency, the T-8 lamp and electronic ballast system qualified for the lighting rebate offered by Boston Edison. The rebate reduced the overall cost in labor and materials to wash, relamp and reballast 1,710 fixtures. The fixtures were lamped with 34-watt, energy-saving lamps and they were driven by standard magnetic ballasts. Fixture wattage measured 155 watts and the lamps were operated 5,408 hours per year. Electric rate for this facility was $0.72 per kilowatt hour. Replacing 6,840 34-watt lamps with 5,130 32-watt, T-8 lamps, an ROI for this facility was calculated to be 22.5%. (See Table 22-1.)

Table 22-1
Energy Analysis Done for Hewlitt Packard

Annual Savings
  Lighting
    Total present lighting load for 6,840 - F40/SS   264.71   kW
    Total new lighting load using 5,130 - F032
      W/Electronic   143.64   kW
    Load reduction due to lighting   121.07   kW

    Annual load reduction for 5,408 hrs/yr   654,747 kWh
    Annual savings at $.0720 per kWh $47,141.78

Return on investment
  Total energy savings for 1,815,613 kWh
    saved over lamp life $130,724,16

  Less total investment for 5,130 new lamps
    at a $15.67 investment per lamp $80,387.10
  Net return on investment $50,337.06

  Annualized net return for 2.8 years $18,147.98

  Return on investment   22.50%

  Pay back period for proposed system   20 months

SECTION VIII
LIGHTING SYSTEMS RESEARCH

Chapter 23
Case Study
Warehouse Lighting
Energy Efficiency,
Effectiveness and Economics[1]

N.K.Falk

INTRODUCTION

In a survey of randomly selected industrial plants, it was found that a surprising proportion of floor space under roof was used for non-fabrication activities. Perhaps it is not so surprising if we consider that raw materials, in process goods, finished products, packing materials, maintenance items, safety gear and ready to ships are all integral aspects of the manufacturing process but functionally are warehouse or storage operations.

Since this survey, changes to increasingly greater automation have further reduced the space in which manual operations are performed, and it is therefore likely that the trend towards increasing storage spaces will continue to accelerate in the decades ahead.

In addition to manufacturing related work space, the distribution industry at all levels is also experiencing an alteration of operating patterns that results in an increase of both the required warehouse space, as well as the complexity of tasks performed in such spaces. In all instances, as the manual labor content in warehousing is reduced, the main operating costs become cost of money for investment in facility and equipment and electric energy cost. Of the latter, lighting is a significant burden.

In such areas, H.I.D. (high intensity discharge) sources, including high pressure sodium, metallic halide and mercury vapor

In such areas, H.I.D. (high intensity discharge) sources, including high pressure sodium, metallic halide and mercury vapor lamps, have found wide application for warehouse lighting primarily because of their high lumen per watt efficacy and their relatively small source size which permits the necessary optical control to achieve very specific and desirable light distributions.

The economic benefits of long lamp life, ease of lamp replacement and small fixture size for minimum interference with warehouse operations have also added to the increasing use of H.I.D. for warehouse lighting, particularly for higher mounting heights of 20 feet or more.

These generally encountered long, narrow aisles can best be illuminated with special asymmetric light distributions that are particularly suitable for producing optimum vertical footcandles on stack surfaces (vertical illumination).

Such lighting fixtures (Figure 23-1) were developed to direct the light (Figure 23-2) within these confined long, narrow and frequently high spaces to provide uniform vertical (stack) and horizontal (on the floor or on a theoretical plane 30 inches above the floor) illumination.

Figure 23-1. Asymmetric warehouse aisle lighting unit designed specifically to generate vertical footcandles.

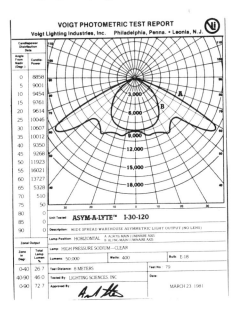

Figure 23-2. Photometric distribution of asymmetric unit shown in Figure 23-1. "A" curve is light pattern along aisle. "B" curve is across aisle.

The performance advantage of employing such light distributions can conveniently be expressed by the use of the concept "spacing ratio" which is the ratio of maximum permissible distance between fixtures along an aisle compared to their mounting height while still providing acceptable uniformity of illumination on critical surfaces (both vertical and horizontal).

With the light pattern generated in Figure 23-2, units can be spaced up to three times as far apart as they are mounted above the floor (Figure 23-3). This 3:1 ratio leads to substantially reduced numbers of luminaires required along aisles (a reduction that could go as high as 50% when compared to conventional highbay units as shown in Figure 23-3a) resulting in a reduction of investment cost and operating expense for items such as lamp maintenance and energy costs.

The economic popularity of H.I.D. special warehouse fixtures has led to their use at lower and lower mounting heights. At these lower applications, down to 14 feet, a number of problems arise. Fortunately, however, solutions have been found in most instances.

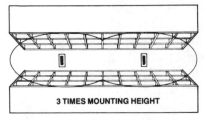

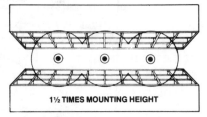

Figure 23-3. Overview of
light provided in an aisle by
asymmetric H.I.D. units
showing maximum allowable
spacing.

Figure 23-3a. Overview of
light provided in an aisle by
highbay H.I.D. units showing
maximum allowable spacing.

Problem

Operators of forklifts, picking cages or automated delivery and
retrieval equipment must look upwards when working on top of the
rack locations. When standing directly below a lighting fixture and
looking upward, a direct view of the lamp arc and reflector causes a
painful and disrupting eye reaction. This has frequently been
sufficiently severe to be an unacceptable condition for virtually any
H.I.D. units at lower and medium heights in warehouse aisles.

Solution

The use of a combination refracting-reflecting optical assembly
(Figure 23-4) used in conjunction with an asymmetric warehouse
aisle lighting distribution producing reflector system.

Figure 23-4.
Asymmetric
warehouse
aisle lighting
unit with a
refractor-
reflector
designed to
limit
brightness to
workers'
eyes.

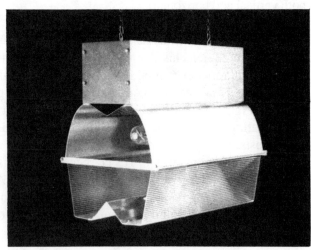

The schematic performance of such an optical device is illustrated (Figure 23-5) which reveals that the solid specular aluminum reflector at the bottom of the refractor-reflector provides total shielding of the lamp for any viewer beneath the luminaire as is illustrated in Fig. 23-6. It is interesting to note that the operator can move to either side of the luminaire, along the aisle, with full lamp obscuration.

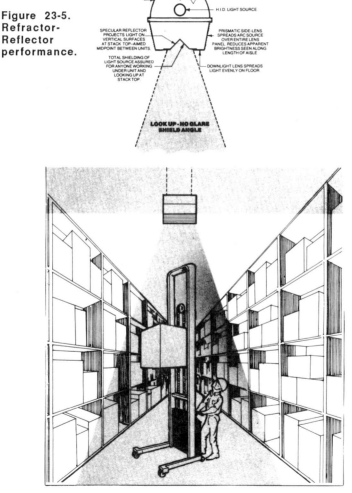

Figure 23-5. Refractor-Reflector performance.

Figure 23-6. Glare protection provided to worker looking at stack top while located under lighting unit.

Problem

Operators, when moving along an aisle on foot or on vehicle, get a "long view" of units that are located at some distance along the aisle and are visible in the upper portion of their normal field of view. Here again, a direct view of the arc and the back wall of the reflector tend to show substantial brightness. And while this brightness is not in the nature of disability glare, it does inflict a real measure of discomfort.

Solution

The prismatic side section of the refractor-reflector tends to split the bright lamp image (both direct and reflected) to reveal virtual multiple source images. This dispersion of source brightness over the entire lens panel area results in significantly lowered lumens per square inch of viewed surface, resulting in a reduction of discomfort glare for observers along the aisle. This can be represented schematically in Figure 23-7.

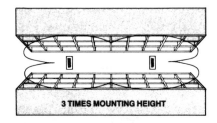

3 TIMES MOUNTING HEIGHT

Figure 23-7. Overview of light provided in an aisle with refractor-reflector asymmetric unit showing reduction of glare normally seen along aisle.

Problem

When materials are stacked very high and close to the ceiling so that the lighting units are mounted at or even below the top of the stored goods, a dark area may occur at a point halfway between fixtures and at the very top of the stacks.

Solution

The reflector in the bottom of the refractor-reflector is also designed to project light onto vertical surface areas at the top of the stacks between luminaire locations. This enhances uniformity of illumination at stack top significantly.

Problem

In some instances, when an insufficient amount of light is reflected back onto the ceiling, a dark ceiling cavity may enhance the brightness contrast between a lighted unit and its dark background, introducing a less than pleasant observer reaction.

Solution

The bottom reflector also redirects light to provide a widespread upward component of 11.12% of total lamp lumen output to relieve the dark ceiling condition.

Problem

The reflector in the reflector-refractor optical lens actually introduces a shadow on the floor directly beneath the unit.

Solution

An additional set of scattering prisms on the bottom surface of the reflector-refractor is employed to fill in any darkened floor area and preserve floor illumination uniformity.

CONCLUSION

Do not space H.I.D. warehouse lighting units closer together than three times their mounting height if you want to save money and energy.

To avoid operator glare when servicing top of the stack goods and for all low mounting height applications, use an appropriate glare shielding mechanism.

References

1. N.K. Falk – "Eye Comfort in Warehouse Situations," Proceedings of the 23rd Annual Meeting of the 1988 IEEE Industrial Application Conference of October, 1988.

2. N.K. Falk – "Warehouse Lighting—It Costs or Pays," Proceedings of the 5th International Conference on Automation in Warehousing – Institute of Industrial Engineers, pp. 217-224, December, 1983.

3. N.K. Falk – "Warehouse Lighting Max/Min Concept," Proceedings of the Industrial Appl. Soc. Meeting, October, 1983.

Chapter 24
Lighting Systems Research*

S.M. Berman, R.R. Verderber, R.D. Clear, C. Greene,
D.D. Hollister, F. Li, D.J. Levy, O.C. Morse,
F.M. Rubinstein, M.J. Siminovitch,
and G.J. Ward

Lawrence Berkeley Laboratory estimates that 50% of the electrical energy consumed by lighting, or about 12% of total national electrical energy sales, could be saved by gradually replacing existing lighting with energy-efficient lighting. This would amount to a yearly savings of some 220 billion kilowatt hours of electricity.

The objective of the Department of Energy's National Lighting Program is to help the lighting community (manufacturers, designers, and users) achieve a more efficient lighting economy. The program, carried out at Lawrence Berkeley Laboratory (LBL), represents a unique partnership between a national laboratory-university complex and industry, facilitating technical advances, strengthening industry capabilities, and providing designers and the public with needed information.

To implement its objectives, the Lighting Program at Lawrence Berkeley Laboratory (LBL) has divided its work into three major categories: advanced light sources, building applications, and impacts on human health and productivity.

The advanced light sources component undertakes research and development projects in lamp technology that are both long-range and high risk. These are projects in which the lighting industry has an interest but which it does not pursue on its own, and from which significant benefits could accrue to both

*This work was supported by the Assistant Secretary for Conservation and Renewable Energy, Office of Buildings and Community Systems, Building Equipment Division of the U.S. Department of Energy under Contract No. DE-AC03-76SF00098.

the public and industry if the technical barriers were surmounted.

The building applications component undertakes research on the electromagnetic compatibility of high-frequency lighting with building functions, including machinery, computers, and other electrical and electronic systems, as well as the interaction of lighting with building energy systems.

The impacts component examines relationships between workers and the physical lighting environment to ensure that energy-efficient technologies contribute to human productivity and health. These efforts are interdisciplinary, involving engineering, optometry, and medicine.

Since its inception in 1976, the LBL Lighting Program has produced more than 118 reports and publications. These reports are available to the public, and document research on subjects such as solid-state ballasts, operation of gas-discharge lamps at high frequency, daylight availability, energy-efficient fixtures, lighting control systems, and visibility and human productivity. The program's internationally recognized interdisciplinary staff spans the fields of engineering, physics, architecture, optometry, and medicine and is involved in a variety of professional, technical, and governmental activities.

The Lighting Program combines the facilities and staff of LBL with those of the University of California College of Environmental Design and School of Optometry, both on the Berkeley campus, and the School of Medicine in San Francisco (UCSF). Because results are directed at enhancing the capabilities and long-term viability of the lighting industry and providing the design profession and the general public with needed information, the program is unique in the United States.

Described below are highlights of the accomplishments realized in FY 1986 by each of our three major efforts—engineering science, building applications, and health impacts—and activities planned for FK 1987. Publications and conference presentations of the past year may be found in the publications list at the end of this chapter.

ADVANCED LIGHT SOURCES

The Advanced Light Sources effort focuses on advanced lamp technology and light source development. To see what can be accomplished in this area, consider that today's fluorescent lamp has a luminous efficacy of approximately 80 lumens of light output per watt of electrical power input. Although this is nearly four times as efficient as an incandescent lamp, still greater efficacies are possible. White light can, theoretically, be produced at almost 400 lumens per watt. The advanced lamp technology program is working to supply the engineering science that will provide the basis for a target efficacy of 200 lumens per watt within the next few years. Table 24-1 lists the series of technical improvements that are the technical elements proposed to reach that goal.

Two significant loss mechanisms have been isolated among the technical targets for achieving more efficient fluorescent lamps: self-absorption of ultraviolet (UV) radiation, and energy loss in lamp phosphors. In the first case, we would like to reduce self-absorption UV radiation, a process that occurs within the lamp plasma before the radiation strikes the phosphor-covered inner wall (the phosphor converts UV radiation into visible light). In the second case, we would like to develop a more efficient phosphor matrix that will convert one energetic UV proton into two visible photons. Reductions in self-absorption could provide a significant improvement, and a two-photon phosphor could double lamp efficacy.

LBL is studying several ways of reducing UV self-absorption. The first method is altering the isotopic composition of mercury. This element has seven stable isotopes, each with slightly different resonance UV emission spectra. Altering the naturally occurring isotopic composition can provide more escape channels for the resonance radiation, thereby reducing the probability of quenching collisions and increasing the amount of UV radiation reaching the phosphor. One possibility for isotope alteration-enrichment with ^{196}Hg is being pursued in a joint effort by LBL and GTE Lighting. Should isotopic alternations prove economical, modified lamps would enter the market

Table 24-1. Targets for improving lighting technologies.

Technology	Comment	Total efficacy (lm/W)	Year entering market
Fluorescent lamps			
High-frequency operation		90	1980
Narrow-band phosphors		100	1983
Isotopically enriched	LBL/DOE	110	1988
Magnetically loaded	technical	120	1990
Two-photon phosphor	initiatives	180	1995
Gigahertz/electrodeless		230	1995
HID Lamps			
Today with (high-freq. ballast)			
400-W high-pressure sodium.		(CRI-25) 100	1984
400-W metal halide		(CRI-66) 80	1984
400-W mercury vapor		50	1984
Electrodeless/high/freq.			
1000-W lamps	10-15% improvement		1989
Low-W lamps	30% improvement		1989
New gases	20-25% improvement		1990
Color-constant/dimmable	20-25% improvement		1993

quickly, as lamps would simply be loaded with isotopically enriched rather than natural mercury, with other lamp manufacturing processes remaining the same.

Another method of reducing UV self-absorption was recently discovered at LBL. It involves an applied magnetic field having a direction parallel to the main current. Axial magnetic field strengths of about 600 gauss can increase light emission by about 6%. LBL and major firms in the lamp industry are studying practical ways to apply this technique.

Reducing the effects of energy loss in the phosphors requires altering a lamp's phosphor material. The materials used today convert each UV photon into, at most, one visible photon.

Improving this conversion rate would increase the efficacy of low-pressure discharge lamps. Although a UV photon has sufficient energy to permit the cascade conversion of the UV photon into two visible photons, this process must occur quickly, and the intermediate level in the cascade must be tuned carefully to ensure that both emitted photons are in the visible spectrum. LBL is examining the possibilities of a program in phosphor chemistry designed to discover whether the two-photon phosphor is feasible. The lamp industry, long aware of the complexity of this problem along with the extensive research effort required to provide solutions, is extremely interested in our results.

A highly promising mechanism developed at LBL uses a plasma coupling principle that allows for lamp plasma excitation to occur primarily near the inner lamp wall thereby reducing the UV transport distance and the likelihood of entrapment loss. This surface wave mode of operation occurs at high frequencies in the radio frequency (RF) range between 100 and 1,000 MHz and permits the lamp excitation without electrodes. This surface wave lamp shows approximately 40% increased energy efficacy over normal fluorescents, operates without starting circuits, and should be very long lasting because of the absence of electrodes.

If these research projects at LBL come to full technological and commercial fruition, future fluorescent lamps should operate at high frequency and be isotopically enriched, magnetically loaded, and coated with a two-photon phosphor. Such lamps would have an efficacy of more than 200 lumens per watt, nearly three times that of today's 60-cycle fluorescent.

Other lighting technology research concentrates on high-intensity discharge (HID) lamps, which could be made both more efficient and dimmable of operated without electrodes. High-frequency operation is required to excite the lamp plasma in an electrodeless mode; it may also permit lamps to function with just one or two metal halides and no mercury or sodium. Electrodeless operation would also enable the use of compounds that have desirable light output and color but that are excluded today because they would harm electrodes.

Finally, an electrodeless lamp that could be dimmed without observable spectral changes and that could provide instant restrike would appeal to lighting designers. Table 23-1 summarizes these goals.

To address the lack of data on the plasma discharges, a program on plasma diagnostics has been initiated to measure electron and ion distribution in both optically thin and thick plasmas.

Mercury Isotope Studies

The photochemical separation process has been identified as the most promising candidate for an economic mercury separation process. Several key steps are crucial to obtaining more information about scavenger processes and scaling up the size of the separation reactor. With this information, better estimates can be made of the engineering and economics of a system viable for the large amounts of mercury isotope needed for lamp application. The problems designated to be studied in FY 1987 are (i) scaling up the mini-reactor to a production rate of a few grams per day, (ii) diagnostic studies ralating to the emission on absorption line shapes for the 3P1 state, (iii) chemical diagnostics to evaluate any reactor processes that lead to undesirable products.

The procedure developed for obtaining a weighed amount of ^{201}Hg isotope in a lamp has been perfected and the ^{201}HG was amalgamated on a gold foil and weighed before and after the Hg was removed by heating. A novel spectroscope procedure has been developed to independently determine the precise amount of added isotope ^{201}Hg. This method does not utilize the diffusion concept which is now in use but will be used to check that methodology. Spectroscopic examination of the hyperfine structure of the green line at 5461A and the blue line at 4046A shows that certain peaks are optically thin and their height proportional to the fractional composition of the ^{201}Hg present in the discharge. A more detailed report and a journal article are in progress.

Magnetic Enhancement

We have developed a number of new test lamps to study the effects of magnetic fields on lamps of different length and diameter. Preliminary results indicate both length and especially diameter cause significant changes in the behavior of magnetic enhancement. General Electric Co. lamp division has agreed to provide us a number of test lamps as a professional courtesy. Some delay is expected in their availability because of scheduling at the GE plant. We expect delivery some time in mid FY 1987.

Surface Wave Lamp

Further improvements have been made in a surface wave launcher with good impedance matching. This new system permits instant starting of the lamp and assures perfect impedance matching once the lamp is ignited, which means that all the RF power is delivered to the plasma with no reflected component— a requirement for improved efficiency. This surface wave launcher caused a 40% gain in efficacy in a standard F15, T-8 fluorescent lamp (15 watt, 1 in. in diameter, 18 in. long), compared to operation at normal power input of 60 Hz. In addition, we have designed and built a new thermal control that provides air movement with extremely tight thermal conditions.

Variable Positive Column Diagnostic

A fundamental problem in surface wave lamp improvements is quantifying the separate contributions of electrode elimination and positive column efficacy. To determine these contributions we have designed and built an electroded lamp that can vary distance between its anode and cathode. The positive column axial dimensions are determined by the use of a precise moving fiber-optic system. As the length between anode and cathode increases, the positive column size increases, and its light output is measured. The end-fall voltages will remain constant during the procedure and thus, by measurements at constant current of total voltage and light output, an extrapolation can be made to determine the end-fall contribution. A more detailed discussion of this topic is being prepared in a technical report and a journal article.

Mercury Absorption Studies

A lamp has been constructed and enriched with [201] Hg from its naturally occurring concentration of 15% to amounts ranging from 30% to 60%. Measurements on the lamp will be made to confirm the Richardson-Berman theory, which predicts that a maximum increase in lamp efficacy will occur in this isotope-concentration range. The results will have significant bearing on the final method used to separate the Hg.

We are planning to complete detailed measurements of the [201] Hg-enriched low-pressure gas-discharge lamp. Measuring the amounts of Hg and isotope additions by both the diffusion method and spectroscopic examination methods will assure the high degree of accuracy critical to selecting the Hg isotope for separation.

The second phase of the photochemical mercury isotope separation program to be carried out by GTE Lighting will begin with the building of the scaled-up reactor.

Magnetic Enhancement

LBL will systematically measure the newly designed lamps housed in the UV integrating cylinder to characterize the phenomenon of magnetic enchancement. The effect will be studied as a function of field intensity, lamp-wall temperature, gas fill, and input power, as well as length and diameter. The angular distribution of the ultraviolet radiation will be examined to determine the extent of polarization. Finally, the change in hyperfine structure will be studied with the high-resolution Jarell-Ash spectrometer that has been installed.

Surface Wave Lamp

The movable length lamp experiment will be carried out at both 60 Hertz and high frequency (30 to 50 Kilohertz). Measurements of the light output will be made by a fiber optics system.

We will continue investigating methods of screening lamps to contain electromagnetic radiation. Initial measurements indicate that the levels at a distance of one foot from the lamps are negligible.

Electrodeless HID Lamp

The initial goal of this project was to develop a low-wattage, efficient HID lamp to replace the incandescent lamp. The effort to date has examined high-wattage, 400-watt lamps. A model of an electrodeless mercury-discharge HID lamp is under development to determine the minimum wattage lamp feasible.

Diagnostics

The Lighting Program has developed a new technique to view and study slices of plasma that are both optically thin and thick. The data can be analytically unfolded to determine the spatial distribution of mercury in the excited state through the plasma. A technical paper on this subject is scheduled to appear in the Journal of Quantitative Spectroscopy this year. A special precision-stepping motor mounted on an optical bench is required for scanning the plasma, and will be designed and constructed for conducting the measurements.

BUILDING APPLICATIONS

This component of the Lighting Program considers technical design mechanisms for assuring the incorporation of energy-efficient lighting in buildings. Specific applications must take into account building types as well as user activities. Building applications also considers the manner in which the lighting system affects other building apparatus, e.g, the HVAC system and indoor air quality. The activities of highest priority include dynamic lighting design (based on the use of automatic lighting controls), the thermal characterization of luminaires, and the use of Peltier devices to prevent lamp thermal losses.

Advanced Lighting Design

A technique has been developed to correctly position photo-cells to sense the ambient illumination level when using daylighting for interior illumination. Photocells should be directed toward the task and measure the illuminance from the task and a suitable large area around the task. The photocell should be shielded so that it does not "see" any portion of the window that is admitting daylight. In addition, the electric lights should be calibrated when there is no daylight (to determine their maximum light output), and during the day (to adjust their proportional response to the amount of illumination the photocell is sensing). Experiments have shown that daylight and electric light can provide specified illumination throughout the entire day.

Fixtures

We have employed a Peltier device to control the minimum lamp wall temperature (MLWT) of a fluorescent lamp at its optimum 40°C temperature. At this temperature a fluorescent lamp's light output and efficacy is maximum. Figure 24-1 below shows the light output of a fluorescent lamp in ambient temperatures ranging from 25°C to 50°C. The lamp, MLWT controlled by the Peltier device, maintains its light output over the entire range, while the light output of an uncontrolled lamp decreases by 25% at a 53°C ambient temperature.

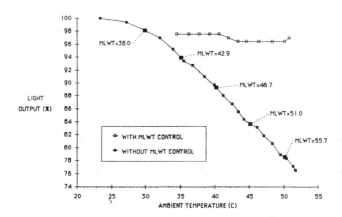

Figure 24-1. Light output vs ambient temperature, with and without Peltier device to control minimum lamp wall temperature. (XBL-8612-5021)

Computer Imaging

A computer imaging technique is being developed for obtaining images of both office and roadway scenes by tracing the path of rays from the source of illumination. The method is an advance over past imaging methods used since actual luminances resulting from both specular and diffuse surfaces can be obtained. This technique permits one to obtain computer images of spaces that appear as realistic as actual photography.

The method has been applied to interior spaces illuminated with daylight as well as electric lights. Recently we have shown the method can provide images of roadways illuminated with lamps and automobile headlights. The technique can be used to show the appearance of pedestrians and roadway signs under different lighting conditions.

Advanced Lighting Design

LBL will develop optimum lighting layouts with computer aided designs, showing the cost effectiveness of static and dynamic designs. The measured performance of various lamp-ballast-fixture systems will be the input to the computer programs, rather than the ideal specifications listed for these products by the manufacturers. Subsequent analysis will include the changes in illumination that occur with the introduction of objects in the space (furniture, etc.). Computer images of the spaces will also be employed to plot the luminances of the surfaces. The analysis will measure the quality of the illumination.

Fixtures

The application of the Peltier device to control the MLWT of fluorescent lamps will be used for the evaluation of fixture parameters. Currently when we measure the efficiency of a fixture, the MLWT of the lamp is different when total light output of the lamp is measured in open air and in the fixture. That is, in the open air the ambient temperature of the lamp will be 22°C to 25°C while in the closed fixture it may be as great as 53°C. To properly assess fixture designs, we need an efficiency metric that is only dependent on the geometric factors of the fixture. We will examine how the Peltier device can be used to obtain this goal.

Computer Imaging

LBL will study the possibility of using computer imaging of scenes to improve visual performance of drivers on the road as well as in a selection of interior workplace environments. The computer program allows one to alter the lighting of spaces so it can be used to evaluate different lighting techniques.

Lighting Controls Demonstration Laboratory

In promoting the use of lighting controls, LBL is attempting to develop a control laboratory to conduct workshops on the proper use of lighting controls. In addition to conducting workshops, this laboratory will do lighting control research and monitor on-site installations on controls.

IMPACTS OF NEW LIGHTING TECHNIQUES ON PRODUCTIVITY AND HEALTH

The idea that lighting might negatively affect health has appeared often in the lay press during the past few years. Scientific data are, however, lacking, especially to ascertain whether new energy-efficient technologies adversely affect human health and productivity.

Performance and productivity may be influenced by the lamp, the electronics and associated controls, the fixture, or the geometry and location of the lighting system. We classify these lighting factors as: color variations; glare; intensity fluctuations; spectrum variations, including the ultraviolet region; electromagnetic fields generated by the lamp, ballast, or controls; and flicker, all of which could evoke a variety of human responses (behavioral, psychophysical, physiological, or biochemical).

Our research seeks to assure that new energy-efficient lighting technologies do not adversely affect human health and productivity. We are investigating whether any aspect of new technologies can produce responses in humans. If we identify responses, we will characterize the effects and identify the necessary changes in lighting technologies. Although subjective responses of workers provide some information, such responses are generally confounded by a mix of sociological factors and individual motivations; the investigations carried out by LBL use objective responses to establish cause and effect and ensure repeatability.

The impacts program is divided into three areas: (1) direct effects of lighting on the human autonomic system (carried out at UCSF and LBL), (2) Interactions, between lighting and visual display terminal operation, that affect productivity or comfort (carried out at the UC School of Optometry and LBL), and (3) analysis of the general relationship between lighting and visual performance (carried out at LBL).

In the first area of this program, lamps to be evaluated include incandescent, cool-white fluorescent, high-pressure sodium, and metal halide. Human responses to various lighting conditions will be assessed by monitoring autonomic responses, including heart rate, galvanic skin response, muscle strength, exercise tolerance, facial expression, and pupillary response. Behavioral measures to be used include memory (Wechsler Memory Scale and Sternberg's Memory Scanning Time), cognitive function (mental arithmetic), time estimation, and simple reaction time. Other behavioral tasks will probably be included.

Data-gathering and subject control are supervised by trained medical personnel. A national technical advisory committee oversees and reviews the project. First results of this effort concern the effects of visible-spectrum and low-frequency radiation on human muscle strength; and as described previously they indicate that subjective psychological factors are the likely cause of reported effects. A second set of experiments using pupillometry has however shown very robust and surprising effects of spectrum on pupil responses to different lamps.

UCSF/LBL Program

LBL has completed experiments using infrared pupillometry to investigate pupil reactions to high-pressure sodium and incandescent lamps. Under conditions of steady state viewing, the results from six subjects have shown a larger pupil diameter from HPS exposure compared with incandescent lamps for the same levels of illumination. This initial result implies that, for visual acuity equivalent to that with incandescent lamps, higher illumination levels will be required with HPS lamps. But, on the other hand, for conditions when acuity is not important, HPS lamps might be used with less illumination. The pupil size differences affected by these lamps is shown in Figure 24-2.

In addition, the potential effect on pupil size of modulation frequency, i.e., 60 Hz and 30 kilohertz, has been studied for four different lamps under sponsorship of LRI (Lighting Research Institute). The absence of any measurable effects has been reported.

Mean & S.D. Across 8 Subjects

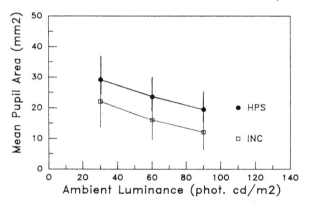

Figure 24-2. The mean pupil area for eight subjects under conditions of equal luminances of incandescent (Inc) and high pressure Sodium (HPS) lamps. (XBL-864-1338)

Visual Performance Program

A model using cost-benefit analysis was applied to the problem of light level recommendations. These recommendations for interior and exterior spaces are currently provided by the Illuminating Engineering Society (IES) on the basis of consensus of expert opinion. Analytical procedures to determine light level recommendations are of concern to the IES and also to those offices in DOE that have responsibilities for proposed building energy use standards and/or guidelines. Cost-benefit analysis results were compared to the IES recommendations. In many cases the results were consistent, but under conditions where the workers were older or the working materials had low reflectivities, recommended light levels were too low. A lengthy report is in progress detailing the specifics of the analysis.

The Effect of Glare on Visual Display Terminals and Other Visual Tasks

Glare from general and task lighting is a concern of both lighting designers and users. One of the most common complaints about lighting results from the perceived presence of glare. The Illuminating Engineering Society provides some guidance on this issue in the Visual Comfort Probability (VCP) concept. Unfortunately this concept has very limited application and is based on experimental work, which, in many cases, is not representative of the modern workplace environment. On the other hand, vision scientists have been unable to find an objective correlate to the subjective sensation of glare. Thus there is much anecdotal concern about lighting glare, but a considerable gap in knowledge about what functions are affected by it.

One of the few conjectures made by vision scientists is that the pupil of the eye or its musculature is the most likely physiological system where glare effects should be manifest. This conjecture is based on the past thirty years of study by Glenn Fry and his co-workers on pupil hippus (small continuous uncontrolled pupil oscillations) and of claims of reduced glare sensitivity under conditions of mydriasis (drugs used to cause the pupil to be partially ot totally dilated).

LBL has undertaken a number of studies carefully examining the frequency spectrum of hippus under a very wide range of illumination conditions using Fourier analysis and methods of infrared pupillometry. Preliminary findings show lack of any dominant frequencies, in contradiction to the results of Fry and others. Furthermore, under conditions of discomfort glare, no apparent changes occurred in the hippus spectrum other than those related to the smooth reduction in oscillation swing observed with continued pupil contraction. Further preliminary results found no support for previous conjectures on the reductions in glare sensation under mydriasis. These studies are continuing and are expected to be completed during FY 1987.

UCSF/LBL Program

On the basis of our work on pupil size under HPS and incandescent lamps, we believe that, under illumination conditions common indoors, the subjective sensation of brightness is correlated with pupil size. Thus it is possible to have a comparison between two different light sources both providing the same luminance level and color appearance (metameric sources), but each producing different pupil sizes. This is possible because we have found a means of producing metameric sources with very different ratios of scotopic to photopic luminance. Under these conditions we can study whether equal luminances and color but different pupil sizes elicit different perceptions of brightness. The predicted response is that the light source with the lower illuminance but larger S/P ratio (smaller pupil) will appear brighter, which is counterintuitive. Several studies along these lines will be undertaken during FY 1987. The first study will use four different fluorescent lamps to create the metameric but differing S.

We will also begin to examine the ways in which pupil size differences induced by lamp spectra affect certain visual functions. The first study will examine the effect of different pupil size (with equal luminance) on contrast sensitivity. The contrast sensitivity stimulus will be provided by a specially designed video display that can present a full range of spacial frequencies both in steady-state and time-varying conditions.

Visual Performance Program

LBL will continue work on cost-benefit methods applied to light level recommendations; the first report will be completed for presentation at the summer 1987 meeting of the IES.

A luminance mapper for rapidly and accurately determining luminances in interior environments will be designed, and a first prototype may be constructed depending on availability of components from the video industry.

A strawman research program will be developed to provide information that will help us objectively determine appropriate light levels in the work place. This effort will be coordinated with all the various members of the lighting community that are presently considering this question, such as NEMA (National Electrical Manufacturers Association), LRI (Lighting Research Institute), IES (Illuminating Engineering Society), ASHRAE, etc.

UC/LBL Optometry Program

Work on the relationship between glare and pupil response will continue. Presentation of efforts and peer review is planned for the April 1987 ARVO (Association for Research in Vision and Opthamology) meeting. Other conditions, such as the effect of differential glare stimuli to each eye, will be examined in terms of their effect on VDT visual performance. We will complete the report summarizing the effects on visual performance of different color letters on VDT screens under both incandescent and fluorescent lighting.

PUBLICATIONS

1. Verderber, R.R. (1986), *Impacts of Daylighting Design Features on the Choice of Lighting Control Systems*, LBL-22043.
2. Richardson, R., and Berman, S. (1986), *Determination of the Excited State Density of an Optically Thick Line.* Presented at the Fourth International Symposium on the Science and Technology of Light Sources, University of Karlsruhe, Germany, April 1986, LBL-21475.
3. Siminovitch, M.J., Rubinstein, F.M., Verderber, R.R., and Clark, T.A. (1986). *The Effects of Fixture Type and HVAC Integration on Fluorescent Lamp/Ballast Performance.* Presented at the IEEE-IAS 1986 Annual Meeting, Denver, CO, LBL-21775.
4. Hollister, D. (1986), "Overview of Advances in Light Sources," To be published in the *Proceedings of the Society of Photo Optical Instrumentation Engineers,* LBL 21820.
5. Li, F., Verderber, R.R., Hollister, D., and Berman, S. (1986), *Improvement of the Efficacy of Fluorescent Lamps by Isotope Blending,* Presented at the Pan Pacific Lighting Exposition, San Francisco, March 1986, LBL-21828.

6. Rubinstein, F.M., Clark, T., Siminovitch, M.J., and Verderber, R.R. (1986), *The Effect of Lighting System Components on Lighting Quality, Energy Use and Life-Cycle Cost.* Presented at the Institute of Electrical and Electronics Engineers, Denver, CO, October 1986, LBL-21884.
7. Berman, S., Jewett, D., Bingham, L., Nahass, R., Perry, F., and Fein, G. (1986), *Pupillary Size Differences Under Incandescent and High Pressure Sodium Lamps.* Presented at the National IES Conference, Boston, MA, August 1986, LBL-21476.
8. Rubinstein, F.M., Ward, G., and Verderber, R.R. (1985), *The Effect of Control Algorithm and Photosensor Response on the Performance of Daylight-Linked Lighting Systems,* LBL-20562.
9. GTE Lighting Products, (1985), *Annual Report: Test, Evaluation, and Report on the Mercury Enrichment for Fluorescent Lamps,* LBL-20614.
10. Jewett, D., Berman, S., Greenberg, M., Fein, G., and Nahass, R. (1985), "Lack of Effects of Human Muscle Strength of the Light Spectrum and Low Frequency Electromagnetic Radiation in Electric Lighting," *Journal of the IES,* November 1985, LBL-20615.
11. Siminovitch, M.J., Rubinstein, F.M., Clark, T., and Verderber, R.R. (1986), *Maintaining Optimum Fluorescent Lamp Performance Under Elevated Temperature Conditions,* Presented at the IES Conference, Boston, MA, August 1986, LBL-21004.
12. GTE Lighting Products (1986), *Physical Diagnostics for Low and High Pressure Gas Discharge Lamps: Final Report,* LBL-22849.

Lighting Systems Research
Building 46, Room 125
Lawrence Berkeley Laboratory
Berkeley, CA 94720

LBL-21004: "Maintaining Optimum Fluorescent Lamp Performance Under Elevated Temperature Conditions," T. Clark, F. Rubinstein, M. Siminovitch, and R. Verderber, April 1986. Presented at the Illuminating Engineering Society Conference, Boston, MA, August 20, 1986.
LBL-21775: "The Effects of Fixture Type and HVAC Integration on Fluorescent Lamp-Ballast Performance," T. Clark, F. Rubinstein, M. Siminovitch, and R. Verderber, June 1986. Presented at the IEEE/IAS 1986 Annual Meeting, Denver, CO, September 29-October 1, 1986.
LBL-22360: "Magnetic Enhancement of Ultraviolet Radiation Efficiency of Low Pressure Hg-Ar Discharge," T. Zhou, L. Wang, D. Hollister, S. Berman and R. Richardson, July 1986. Presented at the Annual IES Conference, Boston, MA, August 16-23, 1986.
LBL-21884: "The Effect of Lighting System Components on Lighting Quality, Energy Use, and Life-Cycle Cost," F. Rubinstein, T. Clark,

M. Siminovitch and R. Verderber, July 1986. Presented at the IEEE Conference, Denver, CO, September 29-October 1, 1986.

LBL-22271: "Building Design: Impact on the Lighting Control System for a Daylighting Strategy," R. Verderber, J. Jewell and O. Morse, October 1986. To be presented at the IEEE/IAS 1987 Annual Meeting, Atlanta, GA, October 19-23, 1987.

Chapter 25

Development of Thermally Efficient Compact Fluorescent Fixtures*

M. Siminovitch, F. Rubinstein, R. Whiteman

INTRODUCTION

This chapter describes the development of thermally efficient compact fluorescent fixtures that employ convective venting to cool lamp wall temperatures. Experimental data show that substantial losses in fixture efficiency can occur as a result of elevated temperatures inside the lamp compartment. These elevated temperatures affect the minimum lamp wall temperature of the compact fluorescent lamp and thus the light output and efficacy of the lamp ballast system.

A series of initial prototype fixtures are described that employ convective venting to reduce elevated temperatures. These cooling strategies can greatly reduce operating lamp temperatures resulting in an increase in both light output and system efficacy.

Compact fluorescent lamps and fixtures are rapidly gaining in application in the lighting of commercial building spaces. These systems are used principally to replace incandescent sources thereby reducing energy use and reducing fixture maintenance. Compact fluorescent lamps are being used in fixture types that have been originally designed for incandescent sources. Optimizing a fixture for the compact fluorescent lamp requires a detailed understanding of both the optical and thermal characteristics of the lamp/fixture

* Presented at 13th World Energy Engineering Congress. Reprinted by permission of The Association of Energy Engineers, Atlanta, GA.

system. The thermal characteristics of the fixture are especially important considerations for fluorescent sources since these are highly sensitive to changes in ambient temperature. Ambient temperature surrounding the lamps within a fixture affects the minimum lamp wall temperature (MLWT) which in turn determines the vapor pressure of the lamp and its output characteristics.

Considerable losses in fixture efficacy can occur as a result of elevated temperatures within the lamp compartment. Typically the geometry of most compact fluorescent fixtures presents a highly constricted thermal environment to the lamps resulting in lamp heating and corresponding light losses that can approach 20%. Efficiency losses that occur due to elevated temperatures can be potentially recovered with careful fixture designs that reduce lamp temperatures. These methods can include heat sinking of the lamp directly or convective cooling of the lamp compartment.

The objective of these studies was to identify the range of light losses that occur due to elevated temperature conditions and to demonstrate the potential of convective venting for enhancing fixture efficiency by reducing lamp temperatures.

Convective venting involves introducing small venting apertures within the geometry of the fixture envelop to promote a natural convection of air through the lamp compartment in order to reduce lamp wall temperatures.

METHODOLOGY

A series of commercial down light fixtures that employ compact fluorescent lamps were examined. For each fixture type, two configurations were tested: the stock fixture without modifications and the same fixture modified to incorporate convective venting through the lamp compartment.

Three generic fixture systems were examined:

1. Recessed down light with two 26-watt compact fluorescent lamps —

This fixture consists of two compact lamps with a horizontal burning position mounted within an open spun aluminum reflector.

Figure 25-1 shows a cross section through the standard fixture with the horizontal burning position of the compact lamps. This fixture was modified to produce a convection directed across the ends of the lamp and then in to the plenum space above. The primary convection pattern is shown on the schematic of the modified fixture.

2. Recessed enclosed down light with two 13-watt compact fluorescent lamps —

This fixture has two 13-watt lamps mounted diagonally within an enclosed lamp compartment. The fixture is recessed mounted within a plenum and has a plastic diffuser flush with the ceiling surface. Figure 25-2 shows a cross section through the vented configuration of the recessed enclosed down light indicating the convection through the fixture to the plenum. The inlet apertures were aligned diametrically and in direct proximity to the ends of the lamps. A single outlet vent was positioned at the top of the fixture.

3. Screw-in 15-watt compact fluorescent lamp/fixture —

This fixture system has a single lamp mounted vertically within a reflector with an Edison base and electronic ballast. Fixture 25-3 shows a cross section of the screw-in fixture within an existing recessed down light and a cross section through the screw-in fixture alone indicating the intended convection pattern. The venting configuration consisted of a series of outlet apertures radially situated around the outside housing of the reflector. These venting apertures allow the heat inside the upper area of the lamp compartment to escape into the existing fixture housing and then into the plenum.

Each fixture system was installed in a test ceiling plane and instrumented with thermistors measuring lamp temperature, ambient temperature and surface temperature of the fixture. Light output and power input characteristics were monitored over time until the lamp/fixture system reached thermal equilibrium. Light output variations for the standard fixture were compared with the variations measured using convective venting approaches. Figure 25-4 shows a cross section of the experimental ceiling plane.

In conjunction with the fixture measurements a study was conducted using a temperature controlled integrating chamber in

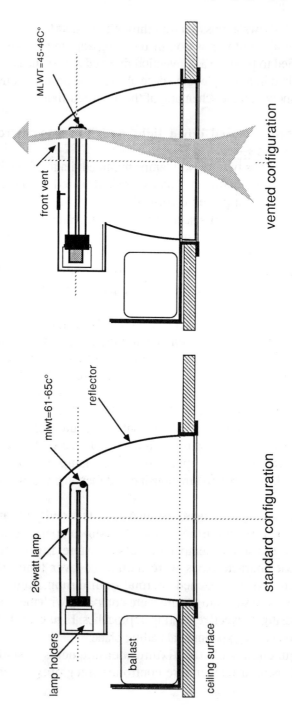

MLWT=45-46C°

front vent

vented configuration

mlwt=61-65c°

reflector

26watt lamp

lamp holders

ballast

ceiling surface

standard configuration

Figure 25-1.
Cross section through recessed down light.

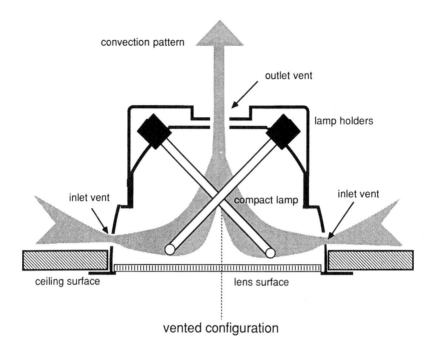

Figure 25-2. Cross section through enclosed down light.

order to characterize the thermal performance of the compact fluorescent lamp operating in bare air. This chamber consisted of a closed rectangular volume which was temperature controlled. A lamp is positioned within the chamber with thermistors attached along its length. A photometer is mounted on the side of the chamber shielded from the direct component of the lamp measuring changes in relative lumen output. Power and ambient temperature are also measured.

The light output and efficacy characteristics of a 26-watt bare lamp were measured over a broad range of ambient temperatures using a commercial 26-watt core coil ballast. The minimum lamp wall temperature (MLWT) was measured at the ends of the lamp farthest away from the base towards the bottom shoulder of the lamp. Figure 25-5 shows a cross section through the temperature controlled integrating chamber.

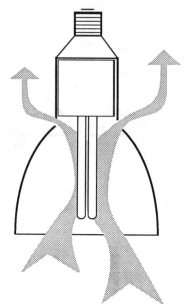

convection through lamp compartment

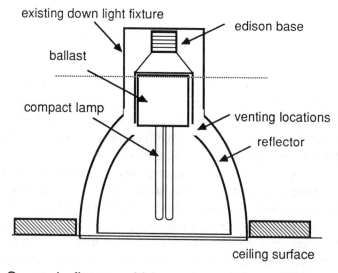

existing down light fixture

edison base

ballast

compact lamp

venting locations

reflector

ceiling surface

Screw-in fixture within recessed down light

Figure 25-3. Retro-fit fixture with recessed down light.

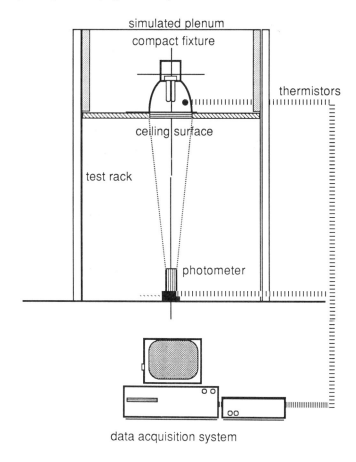

simulated plenum

compact fixture

thermistors

ceiling surface

test rack

photometer

data acquisition system

Figure 25-4. Fixture test station.

EXPERIMENTAL RESULTS

1. Lamp/ballast thermal performance —

Figure 25-6 shows the changes in light output and system efficacy (lamp and ballast) as a function of MLWT for a 26-watt compact fluorescent operating in a horizontal burning position. Light output and system efficacy are maximum at a lamp wall temperature of approximately 40°C with an ambient temperature surrounding the lamps of 27-28°C. As the MLWT exceeds 40°C, light output and system efficacy decrease monotonically. At a MLWT of 60°C, the

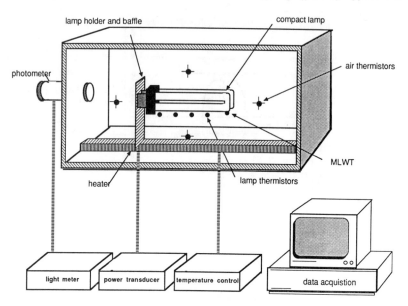

Figure 25-5. Temperature controlled integrating chamber.

light output and efficacy are 20% and 18% below maximum, respectively. It should be noted that this decrease in system efficacy is considerably larger than that encountered in the standard 4-foot fluorescent system for the same change in MLWT.

2.   2x26-watt recessed open down light —

Figure 25-7 shows the change in light output for the stock fixture and the modified recessed 2x26-watt down light as a function of the number of minutes from energizing. For the unmodified fixture, light output quickly reaches a maximum 2-3 minutes after energizing and then asymptotically approaches a minimum of 80% of its maximum over a period of 3 hours. At equilibrium, the MLWT is 60-62°C.

For the fixture modified for convective venting, light output reaches a maximum more slowly (i.e., 4-5 minutes) than the stock fixture, but after reaching equilibrium is only 5% below its maximal value. Thus the venting configuration is shown to effectively cool the lamp so that the MLWT for the vented fixture is only 45-46°C. The vented fixture therefore provides 15% more light than the stock fixture under the same fixture conditions.

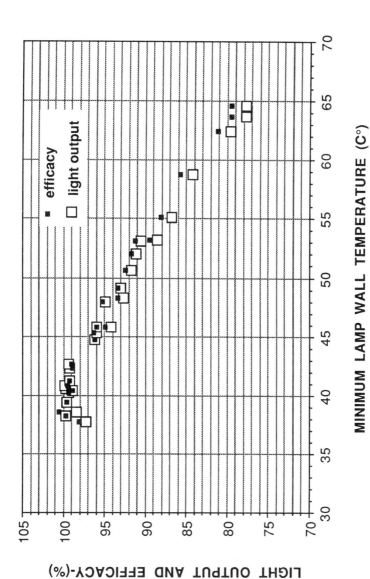

MINIMUM LAMP WALL TEMPERATURE (C°)

Figure 25-6. Light output and efficacy vs. MLWT.

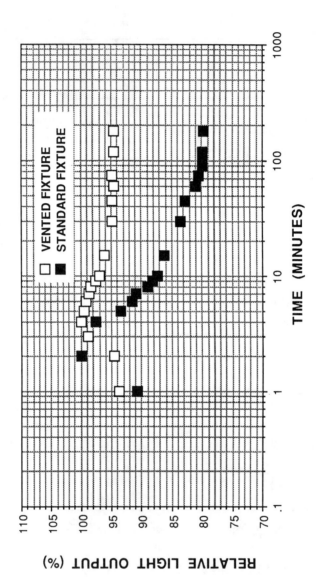

Figure 25-7.
Light output vs. time (open recessed down light).

3. **2x13-watt recessed enclosed down light —**

Figure 25-8 shows the light output variations over time for both configurations of the recessed enclosed down light. As in the open down light discussed above, the light output of the stock fixture drops to 80% of its maximal value after 3 hours' operation.

The fixture with the convective venting maintains near optimum light output over the duration of operation due to the cooler lamp operation. Thus venting the fixture allows the system to produce over 20% more light than the unmodified system.

4. **13-watt screw-in fixtures —**

Figure 25-9 shows the light output variations over time for both the vented and unvented screw-in fixture. For the standard fixture, light output reaches a maximum after 4-5 minutes of operation. Light output stabilizes at approximately 90% of its maximum after a period of 3 hours.

The fixture with convective venting reaches a maximum then reduces to approximately 95% of its maximum. The venting produced 5% increase in light output relative to the standard fixture.

DISCUSSION

The success of convective venting strategies with respect to increased exposure to dirt depreciation is dependent on the application conditions. For the open recessed down light, convective venting significantly increased fixture lumen output and system efficacy compared to the fixture without modifications. This is a good application for venting strategies with respect to dirt depreciation because foreign material entering the fixture compartment will drop out and not be trapped within the compartment. The light loss to the plenum above can be mitigated with appropriately designed light baffles that permit air flow but redirect light back into the fixture.

Convective venting of the enclosed recessed down light also works well in terms of restoring light putout but in this application dirt depreciation may become a serious concern. Apertures in the fixture enclosure will allow insects, dirt and other foreign material to

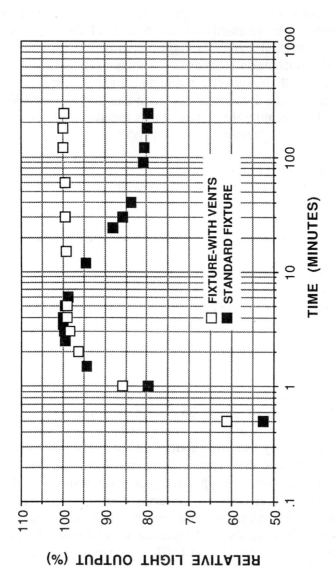

Figure 25-8.
Light output vs. time (enclosed down light).

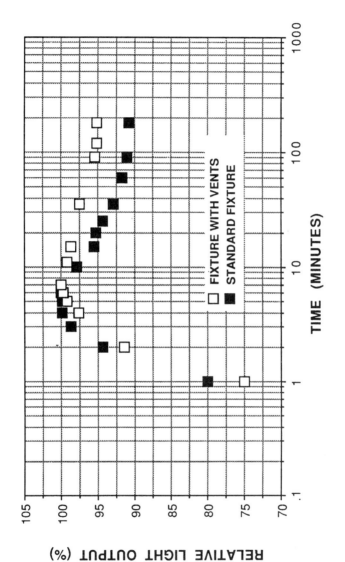

Figure 25-9.
Light output vs. time (screw-in fixture).

enter the fixture. With the enclosed lens surface, material can be trapped inside the fixture. This could result in light losses and potential aesthetic problems.

The application of convective venting in the screw-in retro-fit fixture produced only a 5% increase in light output in comparison to the standard fixture. Venting screw-in or retro-fit fixtures is only a limited utility because the constricted geometry of the existing fixture would restrict the air flow through the compartment. This suggests that heat sinks or other similar methods that conduct heat away from the lamps may be a more appropriate approach to improving thermal performance.

Increasing the lumen output of the lamp/fixture system by reducing lamp temperatures is a critical requirement in order to maximize the application potential of the compact fluorescent lamp in terms of its ability to displace the incandescent light source.

Increased lumen output can potentially expand the market application of compact fluorescents in retro-fit situations by achieving or matching the lumen output of higher wattage incandescents. For example, a 13-watt compact fluorescent lamp with thermal losses may not match the lumen output of a 75-watt incandescent. With a 20% increase in lumen output the 13-watt may approach or surpass the lumen output of a 75-watt lamp watt incandescent. Similar comparisons can be made for all the compact lamp/fixture types ranging from 13 to 26 watts in terms of their ability to achieve incandescent lumen output in a small fixture volume.

Increased lumen output is also significant in new construction. First, increased lumen output of the lamp/fixture system will match higher wattage incandescent fixtures. This will result in furthering the utilization of the compact fluorescent in applications where only incandescent could be used previously. Second, with cooler lamp temperatures the lamp system also shows an increase in system efficacy. In new construction an increase in system efficacy will result in lower power densities and allow fewer fixtures to be used to maintain a specified illuminance level, thus generating a potential savings in building costs.

CONCLUSION

Light losses that can lead to low fixture efficiencies can approach 20% as a function of the constricted thermal environment encountered within compact fluorescent fixtures. Convective venting approaches can significantly reduce lamp temperatures and therefore enhance the light output and efficacy of the fixture system. These approaches appear to be most promising in open recessed down lights with medium to high lumen packages (double 18- or 26-watt lamps).

Major concerns that include light losses to the plenum and U.L. requirements are being examined in the development of advanced prototypes.

ACKNOWLEDGMENT

The work described in this chapter was supported by the Assistant Secretary for Conservation and Renewable Energy, Office of Buildings and Community Systems, Buildings Equipment Division of the U.S. Department of Energy under Contract No. DE-AC03-76SF00098.

Index